T0286428

Textile Effluent Treatment

Textile Effluent Treatment

Edited by **Victor Bonn**

CLANRYE INTERNATIONAL

New Jersey

Published by Clanrye International,
55 Van Reypen Street,
Jersey City, NJ 07306, USA
www.clanryeinternational.com

Textile Effluent Treatment
Edited by Victor Bonn

International Standard Book Number: 978-1-63240-486-2 (Hardback)

Contents

Permissions

List of Contributors

Preface

This book covers various aspects related to the treatment of textile effluents. The management of textile wet processing effluent to meet severe legislative regulations is a complex and constantly changing procedure. Treatment techniques that were acceptable in the past may not be applicable today or in the future. This book presents some concepts and processes to help the textile industry in meeting the demanding requirements of handling textile waste matter.

The researches compiled throughout the book are authentic and of high quality, combining several disciplines and from very diverse regions from around the world. Drawing on the contributions of many researchers from diverse countries, the book's objective is to provide the readers with the latest achievements in the area of research. This book will surely be a source of knowledge to all interested and researching the field.

In the end, I would like to express my deep sense of gratitude to all the authors for meeting the set deadlines in completing and submitting their research chapters. I would also like to thank the publisher for the support offered to us throughout the course of the book. Finally, I extend my sincere thanks to my family for being a constant source of inspiration and encouragement.

Editor

Decolorisation of Textile Dyeing Effluents Using Advanced Oxidation Processes

Taner Yonar

Uludag University, Environmental Engineering Department, Gorukle, Bursa, Turkey

1. Introduction

Textile industry is a leading industry for most countries, such as China, Singapore, UK, Bangladesh, Italy, Turkey etc. But, environmental pollution is one of the main results of this industry. Parralel to usage of huge amounts of water ad chemicals, the textile dyeing and finishing industry is one of the major polluters among industrial sectors, in the scope of volume and the chemical composition of the discharged effluent (Pagga & Brown, 1986).

Textile industry effluents can be classified as dangerous for receiving waters, which commonly contains high concentrations of recalcitrant organic and inorganic chemicals and are characterised by high chemical oxygen demand (COD) and total organic carbon (TOC), high amounts of surfactants, dissolved solids, fluctuating temperature and pH, possibly heavy metals (e.g. Cu, Cr, Ni) and strong colour (Grau, 1991, Akal Solmaz et al., 2006).

The presence of organic contaminants such as dyes, surfactants, pesticides, etc. in the hydrosphere is of particular concern for the freshwater, coastal, and marine environments because of their nonbiodegradability and potential carcinogenic nature of the majority of these compounds (Demirbas at al., 2002, Fang et al., 2004, Bulut & Aydin, 2006, Mahmoudi & Arami, 2006, Mahmoudi & Arami, 2008, Mozia at al., 2008, Li et al., 2008, Atchariyawut et al., 2009, Mahmoudi & Arami, 2009a, Mahmoudi & Arami, 2009b, Mahmoudi& Arami, 2010, Amini et al., 2011,). The major concern with colour is its aesthethic character at the point of discharge with respect to the visibility of the receiving waters (Slokar & Le Marechal, 1997).

The main reason of colour in textile industry effluent is the usage of large amounts of dyestuffs during the dyeing stages of the textile-manufacturing process (O'neil et al., 1999, Georgiou et al, 2002). Inefficient dyeing processes often result in significant dye residuals being presented in the final dyehouse effluent in hydrolised or unfixed forms (Yonar et al., 2005). Apart from the aesthetic problems relating to coloured effluent, dyes also strongly absorb sunlight, thus impeding the photosynthetic activity of aquatic plants and seriously threatening the whole ecosystem. Stricter regulatory requirements along with an increased public demand for colour-free effluent nessesitate the inclusion of a decolorisation step in wastewater treatment plants (Kuo, 1992).

Well known and widely applied treatment method for the treatment of textile industry wastewater is activated sludge process and it's modifications. Combinations of activated sludge process with physical and chemical processes can be found in most applications. These traditional treatment methods require too many spaces and are affected by wastewater flow and characteristic variations. But, either activated sludge process modifications itself or combinations of this process with physical or chemical processes are inefficient for the treatment of coloured waste streams (Venceslau et al., 1994, Willmott et al., 1998, Vendevivere et al., 1998, Uygur & Kok, 1999).

On the other hand, existing physico-chemical advanced treatment technologies such as, membrane processes, ion exchange, activated carbon adsorption etc. can only transfer pollutants from one phase the other phase rather than eliminating the pollutants from effluent body. Recovery and reuse of certain and valuable chemical compounds present in the effluent is currently under investigation of most scientists (Erswell et al., 2002). At this point, The AOPs show specific advantages over conventional treatment alternatives because they can eliminate non-biodegradable organic components and avoid the need to dispose of residual sludge. Advanced Oxidation Processes (AOPs) based on the generation of very reactive and oxidizing free radicals, especially hydroxyl radicals, have been used with an increasing interest due to the their high oxidant power (Kestioglu et al., 2005). In this chapter, discussion and examples of colour removal from textile effluent will be focused on those of most used AOPs.

2. Advanced Oxidation Processes: Principles and definitions

Advanced Oxidation Processes (AOPs) are defined as the processes which involve generation and use of powerfull but relatively non-selective hydroxyl radicals in sufficient quantities to be able to oxidize majority of the complex chemicals present in the effluent water (Gogate & Pandit, 2004a, EPA, 1998). Hydroxyl radicals (OH·) has the highest oxidation potential (Oxidation potential, E_0: 2.8 eV vs normal hydrogen electrode (NHE)) after fluorine radical. Fluorine, the strongest oxidant (Oxidation potential, E_0: 3.06 V) cannot be used for wastewater treatment because of its high toxicity. From these reasons, generation of hydroxyl radical including AOPs have gained the attention of most scientists and technology developers.

The main and short mechanism of AOPs can be defined in two steps: (a) the generation of hydroxyl radicals, (b) oxidative reaction of these radicals with molecues (Azbar et al., 2005). AOPs can convert the dissolved organic pollutants to CO_2 and H_2O. The generation of highly effective hydroxyl radical might possibly be by the use of UV, UV/O_3, UV/H_2O_2, Fe^{+2}/H_2O_2, TiO_2/H_2O_2 and a number of other processes (Mandal et al., 2004).

AOPs can be classified in two groups: (1) Non-photochemical AOPs, (2) Photochemical AOPs. Non-photochemical AOPs include cavitation, Fenton and Fenton-like processes, ozonation at high pH, ozone/hydrogen peroxide, wet air oxidation etc. Short description of some important AOPs are given below. Photochemical oxidation processes include homegenous (vacuum UV photolysis, UV/hydrogen peroxide, UV/ozone, UV/ozone/hydrogen peroxide, photo-Fenton etc), and heterogeneous (photocatalysis etc) processes.

2.1 Non-photochemical oxidation processes

Non-photochemical oxidation processes can be classified as (1) Ozonation, (2) Ozone/Hydregen Peroxide, (3) Fenton Process, (4) Electrochemical Oxidation, (5) Supercritical water oxidation, (6) Cavitataion, (7) Elelctrical discharge-based nonthermal plasma, (8) gamma-ray, (9) x-ray and (10) electron beam. Ozonation, ozone/hydrogen peroxide and Fenton-process are widely applied and examined processes for the treatment of textile effluent. From this reason, brief explanations and examples are given below.

2.1.1 Ozonation

Ozone is well known and widely applied strong oxidizing agent for the treatment of both water and wastewater, in literature and on site. Ozone has high efficiency at high pH levels. At these high pH values (>11.0), ozone reacts almost indiscriminately with all organic and inorganic compounds present in the reacting medium (Steahelin & Hoigne, 1982). Ozone reacts with wastewater compounds in two different ways namely direct molecular and indirect radical type chain reactions. Both reactions occur simultaneously and hence reaction kinetics strongly depend on the characteristics of the treated wastewater (e.g. pH, concentrations of initiators, promoters and scavengers (Arslan & Balcioglu, 2000). Simplified reaction mechanisms of ozone at high pH is given in below;

$$3O_3 + H_2O \xrightarrow{\text{OH}^-} 2OH\bullet + 4 O_2 \tag{1}$$

2.1.2 Ozone/hydrogen peroxide (peroxone) process (O_3/H_2O_2)

The combination of ozone and hydrogen peroxide is used essentially for the contaminants which oxidation is difficult and consumes large amounts of oxidant. Because of the high cost of ozone generation, this combination make the process economically feasible (Mokrini et al., 1997). The capability of ozone to oxidise various pollutants by direct attack on the different bonds (C=C bond (Stowell & Jensen, 1991), aromatic rings (Andreozzi et a. 1991) is further enhanced in the presence of H_2O_2 due to the generation of highly reactive hydroxyl radicals ($\bullet$OH). The dissociation of H_2O_2 results in the formation of hydroperoxide ion, which attacks the ozone molecule resulting in the formation of hydroxyl radicals (Forni et al., 1982, Steahelin & Hoigne, 1985, Arslan & Balcioglu, 2000). General mechanism of peroxon process is given below:

$$H_2O_2 + 2O_3 \rightarrow 2 OH\bullet + 3 O_2 \tag{2}$$

The pH of solution is also critical for the processs efficiency like other AOPs. Addition of hydrogen peroxide to the aqueous O_3 solution at high pH conditions will result in higher production rates of hydroxyl radicals (Glaze & Kang, 1989). Indipendence of peroxone process from any light source or UV radiation gives a specific advantage to this process that it can be used in turbid or dark waters.

2.1.3 Fenton process

The dark reaction of ferrous iron (Fe(ll)) with H_2O_2 known as Fenton's reaction (Fenton 1894), which is shown in Eq.-15, has been known for over a century (EPA, 2001).

$$Fe^{+2} + H_2O_2 \rightarrow Fe^{+3} + OH^- + OH\bullet \tag{3}$$

The hydroxyl radical thus formed can react with Fe(II) to produce ferric ion (Fe(III)) as shown in Eq.-16;

$$\cdot OH + Fe^{+2} \rightarrow Fe^{+3} + OH^- \tag{4}$$

Alternatively, hydroxyl radicals can react with and initiate oxidation of organic pollutants in a waste stream,

$$RH + \cdot OH \rightarrow R\cdot + H_2O \tag{5}$$

At value of pH (2.7–2.8), reactions can result into the reduction of Fe^{+3} to Fe^{+2} (Fenton-like).

$$Fe^{+2} + H_2O_2 \Leftrightarrow H^+ + FeOOH^{+2} \tag{6}$$

$$FeOOH^{+2} \rightarrow HO_2\bullet + Fe^{+2} \tag{7}$$

proceeding at an appreciable rate. In these conditions, iron can be considered as a real catalyst (Andreozziet al., 1991).

At pH values <4.0, ferrous ions decompose H_2O_2 catalytically yielding hydroxyl radicals most directly. However, at pH values higher than 4.0, ferrous ions easily form ferric ions, which have a tendency to produce ferric hydroxo complexes. H_2O_2 is quite unstable and easily decomposes itself at alkaline pH (Kuo, 1992).

Fenton process is cost-effective, easy to apply and effective for the degradation of a wide range of organic compounds. One of the advantages of Fenton's reagent is that no energy input is necessary to activate hydrogen peroxide. Therefore, this method offers a cost-effective source of hydroxyl radicals, using easy-to-handle reagents (Bautista et al., 2007). The Fenton process consisits of four stages. At first, pH is adjusted to low pH. Then the main oxidation reactions take place at pH values of 3-5. The wastewater is then neutralized at pH of 7-8, and, finally, precipitation occurs (Bigda, 1995, Lee & Shoda, 2008). Furthermore, it commonly requires a relatively short reaction time compared with other AOPs. Thus, Fenton's reagent is frequently used when a high reduction of COD is required (Bigda, 1995, Bautista et al., 2007, Lee & Shoda, 2008, Yonar, 2010).

2.2 Photochemical oxidation processes
2.2.1 Homogeneous photochemical oxidation processes
2.2.1.1 Vacuum UV (VUV) photolysis

The Vacuum Ultraviolet range is absorbed by almost all substances (including water and air). Thus it can only be transmitted in a vacuum. The absorption of a VUV photon causes one or more bond breaks. For example, water is dissociated according to;

$$H_2O + h\upsilon(< 190 \text{ nm}) \rightarrow H\bullet + HO\bullet \tag{8}$$

$$H_2O + h\upsilon(< 190 \text{ nm}) \rightarrow H^+ + e^- + HO\bullet \tag{9}$$

Photochemistry in the vacuum-ultraviolet (VUV) spectral domain (approx. 140–200 nm) is of high applicatory interest, e.g. (i) in microelectronics, where materials with surface structures of high spatial resolution provide a basis for the fast development of high computational and electronic and optical storage capacities or (ii) in environmental techniques, in particular for the production of ultra pure water and for the oxidative treatment of waste gas and water (Bolton, 2002, Gonzaleza et al., 2004). VUV-photolysis can be achived by the usage of either a monochromatic (Xe-eximer Xe_2^*) or polychromatic (Hg) radiation sources. Theses light sources have some limitations such as high price, wave length variations etc. From these reasons application of VUV photolysis are too limited.

2.2.1.2 Hydrogen peroxide/UV (H_2O_2/UV) process

This method is based on the direct photolysis of hydrogen peroxide molecule by a radiation with a wavelength between 200-300 nm region. The main reaction of H_2O_2/UV is given below:

$$H_2O_2 + h\upsilon \rightarrow 2\ HO\bullet \qquad (10)$$

The low, medium ad high pressure mercury vapor lamps can be used for this process because it has significant emittance within 220-260 nm, which is the primary absorption band for hydrogen peroxide. Most of UV light can also be absorbed by water. Low pressure mercury vapour lambs usage can lead to usage of high concentrations of H_2O_2 for the generation of sufficient hydroxyl radical. However, high concentrations of H_2O_2 may scavenge the hydoxyle radical, making the H_2O_2/UV process less effective. Some more variables such as temperature, pH, concentration of H_2O_2, and presence of scavengers affect the production of hydroxyl radicals (EPA, 1998, Bolton, 2001, Mandal et al., 2004 Azbar et al., 2005).

2.2.1.3 Ozone/UV (O_3/UV) process

Photolysis of ozone in water with UV radiation in the range of 200-280 nm can lead to yield of hydrogen peroxide. Hydroxyl radicals can be generated by these produced hydrogen peroxide under UV radiation and/or ozone as given equations below:

$$O_3 + h\upsilon + H_2O \rightarrow H_2O_2 + O_2 \qquad (11)$$

$$H_2O_2 + h\upsilon \rightarrow 2\ \cdot OH \qquad (12)$$

$$2O_3 + H_2O_2 \rightarrow 2\ \cdot OH + 3O_2 \qquad (13)$$

Starting from low pressure mercury vapour lamps all kind of UV light sources can be used for this process. Because, O_3/UV process does not have same limitations of H_2O_2/UV process. Low pressure mercury vapor UV lamps are the most common sources of UV irradation used for this process. Many variables such as pH, temperature, scavengers in the influent, tubidity, UV intensity, lamp spectral characteristics and pollutant type(s) affect the effciency of the system (EPA, 1998, Azbar, 2005). Number of laboratory, pilot and full scale applications of Ozone/UV and Hyrdogen peroxide/UV processes can be found in literature. Commercial applications of these processes can also be available.

2.2.1.4 Ozone/hydrogen peroxide/UV (O_3/ H_2O_2/ UV) process

This method is considered to be the most effective and powerful method which provides a fast and complete mineralisation of pollutants (Azbar, 2005, Mokrini et al., 1997). Similar to other ozone including AOPs, increasing of pH affects the hydroyle radical formation. Additional usage of UV radiation also affects the hydroyle radical formation. Efficiency of ozone/hydrogen peroxide/UV process is being much more higher with addition of hydrogen peroxide (Horsch, 2000, Contreras et al., 2001). Main short mechanism of O_3/ H_2O_2/ UV process is given below:

$$2\ O_3 + H_2O_2 \xrightarrow{\ UV\ } 2\ HO\cdot + 3\ O_2 \qquad (14)$$

2.2.1.5 Photo-Fenton process

The combination of Fenton process with UV light, the so-called photo-Fenton reaction, had been shown to enhance the efficiency of Fenton process. Some reasearchers also attributed this to the decomposition of the photo active $Fe(OH)^{+2}$ which lead to the addition of the HO·radicals (Sun & Pignatello, 1993, He & Lei, 2004). The short mechanism of photo-Fenton reaction is given below:

$$Fe(OH)^{+2} + h\upsilon \rightarrow Fe^{+3} + HO\cdot \qquad (15)$$

With $Fe(OH)^{2+}$ being the dominant $Fe(III)$ species in solution at pH 2-3. High valence Fe intermediates formed through the absorption of visible light by the complex between $Fe(II)$ and H_2O_2 are believed to enhance the reaction rate of oxidation production (Pignatello, 1992, Bossmann et al., 2001).

2.2.2 Heterogeneous Photochemical Oxidation processes

Widely investigated and applied Heterogeneous Photochemical Oxidation processes are semiconductor-sentized photochemical oxidation processes.

Semiconductors are characterized by two separate energy bands: a low energy valence band (h^+_{VB}) and a high-energy conduction (e^-_{CB}) band. Each band consists of a spectrum of energy levels in which electrons can reside. The separation between energy levels within each energy band is small, and they essentially form a continuous spectrum. The energy separation between the valence and conduction bands is called the band gap and consists of energy levels in which electrons cannot reside. Light, a source of energy, can be used to excite an electron from the valence band into the conduction band. When an electron in the valence band absorbs a photon,' the absorption of the photon increases the energy of the electron and enables the electron to move into one of the unoccupied energy levels of the conduction band (EPA, 1998).

Semiconductors that have been used in environmental applications include TiO_2, strontium titanium trioxide, and zinc oxide (ZnO). TiO_2, is generally preferred for use in commercial APO applications because of its high level of photoconductivity, ready availability, low toxicity, and low cost. TiO_2, has three crystalline forms: rutile, anatase, and brookite. Studies indicate that the anatase form provides the highest hydroxyl radical formation rates (Korrmann et al., 1991, EPA, 1998).

The photo-catalyst titanium dioxide (TiO_2) is a wide band gap semiconductor (3.2 eV) and is successfully used as a photo-catalyst for the treatment of organic pollutants (Hsiao et al., 1983, Korrmann et al., 1991, Zahhara, 1999). Briefly, for TiO_2, the photon energy required to overcome the band gap energy and excite an electron from the valence band to the conduction band can be provided by light of a wavelength shorter than 387.5 nm. Simplified reaction mechanisms of TiO_2/UV process is given in following equations (eq. 16- eq. 19).

$$TiO_2 + h\upsilon \rightarrow e^-_{CB} + h^+_{VB} \tag{16}$$

$$H_2O + h^+_{VB} \rightarrow OH\bullet + H^+ \tag{17}$$

$$O_2 + e^-_{CB} \rightarrow O_2\bullet^- \tag{18}$$

$$O_2\bullet^- + H_2O \rightarrow OH\bullet + OH^- + O_2 + HO_2^- \tag{19}$$

The overall result of this reversal is generation of photons or heat instead of -OH. The reversal process significantly decreases the photo-catalytic activity of a semiconductor (EPA, 1998). Main advantage of TiO_2/UV process is low energy consumption which sunlight can be used as a light source.

3. Characterisation of textile industry wastewater

Textile industry produces large amounts of liquid by-products. Volume and composition of these waswater can vary from one source to other source. In the scope of volume and the chemical composition of the discharged effluent, the textile dyeing and finishing industry is one of the major polluters among industrial sectors. Textile industry dyes are intentionally designed to remain photolytically, chemically and biochemically stable, and thus are usually not amenable to biodegradation (Pagga & Braun, 1986). Like many other industrial effluents, textile industry wastewater varies significantly in quantity, but additionally in composition (Correira et al., 1994).

These wastes include both organic and inorganic chemicals, such as finishing agents, carriers, surfactants, sequestering agents, leveling agents etc. From these reasons, textile effluents are characterized with high COD ($\approx$ 400-3.000 mg/L), BOD$_5$ ($\approx$ 200-2.000 mg/L), Total Solids ($\approx$ 1.000-10.000 mg/L), Suspended Solids ($\approx$ 100-1.000 mg/L), TKN ($\approx$ 10-100 mg/L), Total Phosporus ($\approx$ 5-70 mg/L), Conductivity (1.000-15.000 mS/cm) and pH ($\approx$ 5-10 usually basic) (Grau, 1991, Pagga ad Braun, 1991, Kuo, 1992, Correira et al., 1994, Arslan and Balcioglu, 2000, , Nigam et al., 2000, Azbar et al., 2005, Akal Solmaz et al., 2006, Yonar et al., 2006, Mahmoudi & Arami, 2009, Yonar, 2010,).

Another important problem of textile industry wastewater is color. Without proper treatment of coloured wate, these dyes may remain in the environment for a long time (Yonar et al, 2005). The problem of colored effluent has been a major challenge and an integral part of textile effluent treatment as a result of stricter environmental regulations. The presence of dyes in receiving media is easily detectable even when released in small concentrations (Little et al., 1974, Azbar et al., 2004). This is not only unsightly but dyes in the effluent may have a serious inhibitory effect on aquatic ecosystems as mentioned above (Nigam et al., 2000).

Definition and determination of colour is another important point for most water and wastewater samples. Some methods can be found in literature for the determination of colour in samples. But, selection of true method for the determination of colour is very important. According to "Standard Methods for the Examination of Water and Wastewater" (APHA- AWWA, 2000), importance of colour is defined with some sentences given below:

"Colour in water may result from the presence of natural metallic ions (iron and manganese), humus and peat materials, plankton, weeds, and *industrial wastes*. Colour is removed to make a water suitable for general and industrial applications. Coloured industrial wastewaters may require colour removal before discharge into watercourses."

From these reasons, colour content should be determined carefully. In Standard Methods, colour content of water and wastewater samples can be determined with four different methods such as (i) Visual Comparison Method, (ii) Spectrometric Method, (iii) Tristimulus Filter Method, and (iv) ADMI Tristimulus Filter Method. Selection of true and appropriate method for samples is very important. Visual comparison method is suitable for nearly all samples of potable water. This method is also known as Platinum/Cobalt method. Pollution by ceratin industrial wastes may produce unusual colour that can not be easily matched. In this case, usage of instrumental methos are appropriate for most cases. A modification of the spectrometric and tristimulus methods allows calculation of a single colour value representing uniform cromaticity differences even when the sample exhibits colour significantly different from that of platinum cobalt standards (APHA-AWWA, 2000).

4. Colour removal from textile industry wastewater by AOPs

Most commonly applied treatment flow scheme for textile effluent in Turkey and other countries generally include either a single activated sludge type aerobic biological treatment or combination of chemical coagulation and flocculation + activated sludge process (Yonar et al., 2006). Furthermore, it is well known that aerobic biological treatment option is ineffective removal for colour removal from textile wastewater in most cases and the chemical coagulation and flocculation is also not effective for the removal of soluble reactive dyestuffs. Therefore, dyes and chemicals using in textile industry in effluent may have a serious inhibitory effect on aquatic ecosystems and visual pollution on receiving waters, as mentioned above (Venceslau et al., 1994, Willmott et al., 1998, Vendevivere et al., 1998).

There are several alternative methods used to decolorize the textile wastewater such as various combinations of physical, chemical and biological treatment and colour removal methods, but they cannot be effectively applied for all dyes and these integrated treatment methods are not cost effective. Advanced Oxidation Processes (AOPs) for the degradation of non-biodegradable organic contaminants in industrial effluents are attractive alternatives to conventional treatment methods and are capable of reducing recalcitrant wastewater loads from textile dyeing and finishing effluents (Galindo et al., 2001, Robinson et al., 2001, Azbar et al., 2004, Neamtu et al., 2004). In this section, applied AOPs for colour removal from textile effluent are given. Technological advantages and limitations of these AOPs is also discussed.

4.1 Colour removal with non-photochemical AOPs

Ozonation at high pH, ozone/hydrogen peroxide and Fenton processes are widely applied and investigated AOPs for colour removal from textile effluents and tetile dyes. As it can be clearly seen from former sections, ozone can produce hydroxyl radicals at high pH levels. According to this situation, pH is very important parameter for ozonation process. As it was described above, under conditions aiming hydroxyl free radical (HO•) production (e.g., high pH), the more powerful hydroxyl oxidation starts to dominate (Hoigne & Bader, 1983). Since the oxidation potential of ozone reportedly decreases from 2.07mV (acidic pH) to 1.4mV (basic pH) (Muthukumanar et al., 2001), it is clear that another more powerful oxidant (HO•) is responsible for the increase in the dye degradation, with a consequent colour absorbance decrease. The efficiency of ozonation in the removal of colourand COD from textile wastewater is important to achieve to discharge limits (Somensia et al., 2010).

Textile wastewaters is very complex due to the organic chemicals such as many different dyes, carriers, biocides, bleaching agents, complexion agents, ionic and non-ionic surfactants, sizing agents, etc. As a result, it is hard to explain the overall degradation of the organic matter by ozone in textile wastewater individually. Thus, some global textile wastewater parameters such as color, COD and dissolved organic carbon are used for the degradation kinetic of organic matter by ozonation (Sevimli & Sarikaya, 2002, Selcuk, 2005).

Textile wastewaters exhibit low BOD to COD ratios (< 0.1) indicating non-biodegradable nature of dyes and Wilmott et al.(1998) have claimed that aerobic biological degradation is not always effective for textile dye contaminated effluent (Sevimli & Sarikaya, 2002).

Somensia et al., (2010) , tested pilot scale ozonation for the pre-treatment and colour removal of real textile effluent. Authors have mentioned that the importance of pH on the process efficiency and colour removal efficiencies were determined as 40.6% and 67.5% at pH 3.0 and 9.1, respectively. COD removal effcieincies ware also determined as 18.7% (pH=3) and 25.5% (pH=9). On the other hand, toxicity can be reduced significantly compared with raw wastewater. Azbar et al., (2004) carried out a comparative study on colourand COD removal from acetate and fiber dyeing effluent. In this study, various advanced oxidation processes (O_3, O_3/UV, H_2O_2/UV, $O_3/H_2O_2/UV$, Fe^{+2}/H_2O_2) and chemical treatment methods using $Al_2(SO_4)_3.18H_2O$ $FeCl_3$ and $FeSO_4$ for the Chemical Oxygen Demand (COD) and colour removal from a polyester and acetate fiber dyeing effluent is undertaken. Ozonation showed superior performance at pH=9 and 90% COD and 92% colour can be removed. Akal Solmaz et al., 2006, applied ozonation to real textile wastewaters and found 43% COD and 97% colour removal efficiencies at pH 9 and C_{O3} 1.4 g/h. In the another study of Akal Solmaz et al., (2009), group has tested different AOPs on two different textile wastewater. 54-70% COD removal and 94-96 % colour removal efficiencies have been determined at pH = 9.

In another study, Selcuk, (2005), have tested coagulation and ozonation for color, COD and toxicity removal from textile wastewater. Author found that, ozonation was relatively effective in reducing colour absorbances and toxic effects of textile effluents compared with chemical coagulation. Almost complete colourabsorbances (over 98%) were removed in 20 min ozone contact time, while COD removal (37%) was very low and almost stable in 30 min ozonation period.

Yonar et al., (2005), have been studied AOPs for the improvement of effluent quality of a textile industry wastewater treatment plant. Authors were mainly tested homogeneous photochemical oxidation processes (HPOP's) (H_2O_2/UV, O_3/UV and $H_2O_2/O_3/UV$) for colour and COD removal from an existing textile industry wastewater treatment plant effluent together with their operating costs. At pH=9, 81% COD and 97% colour removal efficiencies were reported for ozonation process.

As it can be clearly seen from literature, ozonation is very effective for the removal of colour from textile wastewater. COD and toxicity can also be removed by ozonation. But, for decision making on these processes advantages and limitations of these processes should be known. Main advantage of ozonation is no need to addition of any chemicals to water or wastewater. Because, ozone is mostly produced by cold corona discharge genertors. And these generators need dry air for the production of ozone. On the other hand, sludge or simiar residues is not produced during this process. At this point, specific advantage can be stated for textile effluents. Mostly, the pH value of textile wastewater are higher than 7 and in some situations higher than 9. Thus, ozonation can be applied to textile effluent without any pH adjustment and chemical addition. But, ozonation process has some disadvantages, such as, inefficient production capacities of cold corona discharge (CCD) generators (2-4%), less solubility of gas phase ozone in water, higher energy demads of CCD generators, possible emission problems of ozone etc. These disadvantages can be overcomed by the production of efficient ozone generators like membrane electrochemical ozone generators.

Ozone/Hydrogen peroxide process is onother efficient AOPs for the treatment of recalcitrant organics. Similar to ozonation, ozone including other processes mostly needs alkaline conditions. This argument has been extensively and successfully studied by Hoigne (1998) in the attempt of giving a chemical explanation to the short life time of ozone in alkaline solutions. Hoigné showed that the ozone decomposition in aqueous solution develops through the formation of hydroxyl radicals. In the reaction mechanism OH^- ion has the role of initiator:

$$HO^- + O_3 \rightarrow OH_2^- + O_2 \qquad (20)$$

$$OH_2^- + H^+ \Leftrightarrow H_2O_2 \qquad (21)$$

$$OH_2^- + O_3 \rightarrow HO_2\bullet + O_2\bullet^- \tag{22}$$

$$HO_2\bullet \Leftrightarrow H^+ O_2\bullet^- \tag{23}$$

$$O_2\bullet^- + O_3 \rightarrow O_2 + O_3\bullet^- \tag{24}$$

$$O_3\bullet^- + H^+ \rightarrow HO_3\bullet \tag{25}$$

$$HO_3\bullet \rightarrow OH\bullet + O_2 \tag{26}$$

$$OH\bullet + O_2 \rightarrow HO_2\bullet + O_2 \tag{27}$$

It is clear therefore that the addition of hydrogen peroxide to the ozone aqueous solution will enhance the O_3 decomposition with formation of hydroxyl radicals. The influence of pH is also evident, since in the ozone decomposition mechanism the active species is the conjugate base HO_2^- whose concentration is strictly dependent upon pH. The increase of pH and the addition of H_2O_2 to the aqueous O_3 solution will thus result into higher rates of hydroxyl radicals production and the attainment of higher steady-state concentrations of hydroxyl radicals in the radical chain decomposition process (Glaze & Kang, 1989). It must be remarked that the adoption of the H_2O_2/O_3 process does not involve significant changes to the apparatus adopted when only O_3 is used, since it is only necessary to add an H_2O_2 dosing system (Andreozzi, 1998).

Hydrogen peroxide/ozone (peroxone) process test result for real or synthetic textile wastewater are too limited in literature but ozone and hydrogen peroxide is a very promising technique for potential industrial implementation. Kurbus et al. (2003) were conducted comperative study on different vinylsulphone reactive dyes. For all tested dyes, over 99% colour removal can be achieved at pH=12. Kos & Perkovski (2003), were tested different AOPs including peroxone process on real textile wastewater. Textile wastewater initial COD is over 5000 mg/L and authors declared that nearly 100% colour removal can be achived with peroxone process. According to Akal Solmaz et al., (2006), addition of hydrogen peroxide to ozone is increased colour and COD removal efficiencies nearly 10%. Perkovski et al., (2003), were tested peroxone process on anthraquinone dye Acid Blue 62 and they found 60% colour removal efficiency.

Main advantage and disadvantage of peroxone process is addition of hydrogen peroxide. Addition of hydrogen peroxide is giving higher efficiencies and no need to upgrade the existing ozonation systems. But, addition of hydrogen peroxide means additional costs for the treatment of wastewater.

Finally, Fenton process is mostly applied on both textile and other industrial wastewaters. Nevertless, the high electrical energy demand is general disadvantage of most AOPs. As it mentioned above, the greatest advantages of Fenton process is that no energy input is necessary to activate hydrogen peroxide. Most other AOPs need energy input for this activation such as UV based processes, US based proceeses, wet air oxidation etc.

The dark reaction of ferrous ion with hydrogen peroxide was found by Fenton (1894). During the last decades, important scientific studies were carried out on the treatment of most toxic chemicals and waste streamns with this process. Another advantage of Fenton process is the applicability of this process in full scale. Because, this process can be accepted as the modification of traditional physico-chemical treatment. Fenton process can control in different steps of mixing and settling processes. By other words this process does not need specific and complex reactor designs. But, the main important disadvantage of this process among all AOPs is sludge production. Ferric salts should be settled and disposed before discharge of the effluent.

Treatment efficiencies and results of applied Fenton process results in literature summarized in Table 1. According to these results, Fenton process is also promising technique for the treatment and decolorisation of textile effluent.

COD removal (%)	Colour removal (%)	pH	C_{H2O2} (mg/L)	$C_{FeSo4} - C_{FeCl3}$ (mg/L)	Literature
64-71	78-95	3	200-400	200-400	Akal Solmaz et al. (2006)
43-58	92-97	3	100-200	150-200	Akal Solmaz et al. (2009)
84-87	90-91	3-3.5	200-250	200-250	Yonar (2010)
94	96	5	300	500	Azbar et al (2005)
59	89	3.5	800	300	Meriç et al (2005)
29	65	4	70	20	Üstün et al. (2007)
67	90	3-5	150	150	Lin et al. (1997)
93 (TOC)	99	2.45	5200	3600	Liu et al. (2007)
16-22	92-96	4	10	5	Kang et al. (2002)

Table 1. Results of Fenton process in literature in terms of COD and colour removal

4.2 Colour removal with photochemical AOPs

For the treatment and decolorisation of textile effluent, photochemical oxidation processes are widely investigated in literrure. Photochemical oxidation processes are good and emerging alternatives and need UV radiation for the production of hydroxyl radicals. Vakuum UV phooxidation is most powerful member of these processes. Hydroxyl radicals can be produced with VUV with no any chemical addition. Generally Ve-eximer lamp are employed for VUV band radiation. In literature, a number of studies can be found for the treatment of organics with VUV. Despite numerous positive examples, the theory of reactor modelling for sharply nonuniform light distribution is not well developed (Braun et al., 1993). Main reason of this situation is the high price of Xe-eximer lambs.

Tarasov et al., (2003) investigated VUV photolysis for dye oxidation. They tested VUV process on 6 different dye solutions (methylene blue (Basic Blue 9), Basic blue Zn-salt; Direct Green 6; fucsine; Acid Yellow 42, Acid Yellow 11). Degradation of all dyes under VUV condition takes place in about a minute. In another study, Al-Momani et al. (2002) studied photo-degradation and biodegradability of three different families of non-biodegradable textile dyes (Intracron reactive dyes, Direct dyes and Nylanthrene acid dyes) and a textile wastewater, using VUV photolysis. Ninety percent of colour removal of dye solutions and wastewater is achieved within 7 min of irradiation.

UV/H_2O_2 is one of the popular and commercial advanced oxidation process. Like other AOPs, the reaction pH of the treatment system has been observed to significantly affect the degradation of pollutants (Sedlak & Andren, 1991, Lin & Lo, 1997, Kang & Hwang, 2000, Nesheiwat & Swanson, 2000, Benitez et al., 2001a). The optimum pH has been observed to be 3 in the majority of the cases in which H_2O_2 was used with UV irradiation (Ventakandri & Peters, 1993, Tank & Huang, 1996, Kwon et al., 1999, Benitez et al., 2001b) and hence is recommended as the operating pH. It should be noted here that the intrinsic rates of UV/H2O2 process may not be affected much, but at lower operating pH, the effect of the radical scavengers, especially ionic such as carbonate and bicarbonate ions, will be nullified leading to higher overall rates of degradation. Thus, it is better to have lower operating pH (Gogate & Pandit, 2004b).

In literature, hydrogen peroxide (H_2O_2) itself acts as an effective hydroxyl radical (OH·) scavenger at high concentrations given in following empirical equation (Arslan, 2000).

$$H_2O_2 + OH· \rightarrow HO_2· + H_2O \qquad k = 1.2\text{-}4.5\ 10^7\ M^{-1}\ s^{-1} \qquad (28)$$

Although $HO_2·$ promoted radical chain reactions and it is an oxidant itself, its oxidation potential is much lower than that of hydroxyl radical (OH·). Thus, the presence of excess hydrogen peroxide (H_2O_2) can lower the treatment efficiency of AOPs and it is very important to optimize the applied hydrogen peroxide (H_2O_2) concentration to maximize the treatment performance of AOPs (Arslan, 2000).

The presence in the treated water of carbonate can result in in a significant reduction of the efficiency of abetement of pollutants as explained in some studies (Bhattacharjee & Shah, 1998, Andreozzi et al., 1999). Carbonate acts as radical scavenger;

$$HCO_3^- + OH· \rightarrow CO_3^-· + H_2O \qquad k_{HCO3\text{-},\bullet OH} = 1.5\ 10^7\ M^{-1}\ s^{-1} \qquad (29)$$

$$CO_3^{-2} + OH· \rightarrow CO_3^-. + OH^- \qquad k_{CO32\text{-},\bullet OH} = 4.2\ 10^8\ M^{-1}\ s^{-1} \qquad (30)$$

since CO_3^-. is much les reactive than hydroxyl radical (OH·) inhibition by carbonate influences the behavior of most AOPs. At lower operating pH values, the effect of radical scavengers, especially ionic such as carbonate and bicarbonate will be nullified leading to higher overall rates of degradation (Gogate & Pandit, 2004a). Thus, lower operating pH values are recommended for most AOPs in literature. Galindo & Kalt, (1998) documented that the H_2O_2/UV process was more effective in an acid medium (pH ≈ 3-4) in term of discolouration.

On the other hand, the aqueous stream being treated must provide good transmission of UV light, so that turbidity and high suspended solids concentration would not cause interferences. Scavengers and excessive dosages of chemical additives may inhibit the process. Heavy metal ions (higher than 10 mg l⁻¹), insoluble oil and grease, high alkalinity and carbonates may cause fouling of the UV quartz sleeves. Therefore, a good pretreatment of the aqueous stream should be necessary for UV based AOPs (Azbar et al, 2005).

Decolorisation and treatment of textile effluent were investigated in most studies (Shu et al., 1994, Galindo & Kalt, 1998, , Arslan and Balcıoğlu, 1999, Ince, 1999, Neamtu et al., 2002, Cisneros, 2002, Mohey El-Dein et al., 2003, Azbar et al, 2004, Shu & Chang, 2005, Yonar et al. 2005). According to these studies, the use of H_2O_2/UV process seems to show a satisfactory COD (70-95%) and colour(80-95%) removal performance.

According to Rein (2001), conventional ozonation of organic compounds does not completely oxidize organics to CO_2 and H_2O in many cases. Remaining intermediate products in some solution after oxidation may be as toxic as or even more toxic than initial compound and UV radiation could complete the oxidation reaction by supplement the reaction with it. UV lamp must have a maximum radiation output 254 nm for an efficient ozone photolysis. The O_3/UV process is more effective when the compounds of interest can be degraded through the absorption of the UV irradiation as well as through the reaction with hydroxyl radicals (Rein, 2001; Metcalf and Eddy, 2003). The O_3/UV process makes use of UV photons to activate ozone molecules, thereby facilitating the formation of hydroxyl radicals (Al-Kdasi et al., 2004).

Hung-Yee & Ching-Rong (1995) documented O_3/UV as the most effective method for decolorizing of dyes comparing with UV oxidation by UV or ozonation alone. While, Perkowski & Kos (2003) reported no significant difference between ozonation and O_3/UV in terms of colour removal. Even though ozone can be photodecomposed into hydroxyl radicals to improve the degradation of organics, UV light is highly absorbed by dyes and very limited amount of free radical (HO·) can be produced to decompose dyes. Thus same colour removal efficiencies using O_3 and O_3/UV could be expected. In normal cases, ozone itself will absorb UV light, competing with organic compounds for UV energy. However, O_3/UV treatment is recorded to be more effective compared to ozone alone, in terms of COD removal. Bes-Piá et al. (2003) documented that O_3/UV treatment of biologically treated textile wastewater reduced COD from 200-400 mg/L to 50 mg/L in 30 minutes, while, using ozone alone COD reduced to 286 mg/L in same duration. Azbar et al. (2004) documented that using O_3/UV process high COD removal would be achieved under basic conditions (pH=9). Yonar et al. also repoted that using O_3/UV process showed high COD removal efficiency under similar conditions (pH=9) for physically and biologically treated textile effluent.

The addition of H_2O_2 to the O_3/UV process accelerates the decomposition of ozone, which results in an increased rate of OH• generation (Teccommentary, 1996). In literature most AOPs applied for the treatment of textile effluent and, among the all apllied AOPs for dye house wastewater, acetate, polyester fiber dying process effluent and treatment plant outlet of textile industry with the combination of $H_2O_2/O_3/UV$ appeared to be the most efficient in terms of decolouration (Perkowski & Kos, 2003, Azbar et al., 2004, Yonar et al, 2006).

The rate of destruction of organic pollutants and the extent of mineralisation can be considerably increased by using an $Fe(II,III)/H_2O_2$ reagent irradiated with near-UV and/or visible light (Goi & Trapio, 2002, Torrades et al., 2003, Liou et al., 2004, Murugunandham & Swaminathan, 2004), in a reaction that is called the "photo-Fenton reaction". This process involves the hydroxyl radical (HO.) formation in the reaction mixture through photolysis of hydrogen peroxide (H_2O_2/UV) and fenton reaction (H_2O_2/Fe^{+2}.) (Fenton, 1894; Baxendale and Wilson, 1956). Using the photo-fenton process to treat dye manufacturing wastewater, which contains high strength of color, and the results demonstrated its great capability for colour removal (Kang et al., 2000; Liao et al., 1999). Since the hydroxyl radical is the major oxidant of the photofenton process, the removal behavior of COD and colouris highly related with the hydroxyl radical formation. However, the relation between the removal of COD and colourwith the hydroxyl radical formation in the decolorisation of textile wastewater by photo-fenton process was rarely found in the literature.

The colour removal is markedly related with the amount of hydroxyl radical formed. The optimum pH for both the hydroxyl radical formation and colour removal occurs at pH 3±5. Up to 96% of colour can be removed within 30 min under the studied conditions. Due to the photoreduction of ferric ion into ferrous ion, colour resurgence was observed after 30 min. The ferrous dosage and UV power affect the colour removal in a positive way, however, the marginal benefit is less signifcant in the higher range of both (Kang et al., 2000)

Liu et al., 2007 investigated the degradation and decolorisation of direct dye (Everdirect supra turquoise blue FBL), acidic dye (Isolan orange S-RL) and vat dye (Indanthrene red FBB) by Fenton and UV/Fenton processes. A comparative study for Fenton and UV/Fenton reactions by photoreactor has been carried out by scale-up of the optimum conditions, obtained through jar-test experiments. Fenton process is highly efficient for colour removal for three dyes tested and for TOC removal of FBB and FBL. The optimum pH values obtained were all around 3 for FBL, FBB and S-RL. UV/Fenton process improved slightly for FBB and FBL treatment efficiencies compared to Fenton reaction while S-RL showed much better improvement in TOC removal.

The photolysis and photo-catalysis of ferrioxalate in the presence of hydrogen peroxide with UV irradiation (UV/ferrioxalate/H_2O_2 process) for treating the commercial azo dye, reactive Black B (RBB), is examined. An effort is made to decolorize textile effluents at near neutral pH for suitable discharge of waste water. pH value, light source, type of initial catalyst (Fe^{3+} or Fe^{2+}) and concentration of oxalic acid (Ox) strongly affected the RBB removal efficiency. The degradation rate of RBB increased as pH or the wavelength of light declined. The optimal molar ratio of oxalic acid to Fe(III) is three, and complete colour removal is achieved at pH 5 in 2 h of the reaction. Applying oxalate in such a photo process increases both the RBB removal efficiency and the COD removal from 68% and 21% to 99.8% and 71%, respectively (Huang et al., 2007)

Neamtu et al. (2004) investigated the degradation of the Disperse Red 354 azo dye in water in laboratory-scale experiments, using four advanced oxidation processes (AOPs): ozonation, Fenton, UV/H_2O_2, and photo-Fenton. The photodegradation experiments were carried out in a stirred batch photoreactor equipped with an immersed low-pressure mercury lamp as UV source. Besides the conventional parameters, an acute toxicity test with a LUMIStox 300 instrument was conducted and the results were expressed as the percentage inhibition of the luminescence of the bacteria Vibrio fisheri. The results obtained showed that the decolorisation rate was quite different for each oxidation process. After 30 min reaction time the relative order established was: UV/H_2O_2/Fe(II) > Dark/H_2O_2/Fe(II) > UV/H_2O_2/O_3 > UV/H_2O_2/Lyocol. During the same reaction period the relative order for COD removal rate was slightly different: UV/H_2O_2/Fe(II) > Dark/H_2O_2/Fe(II) > UV/H_2O_2 > UV/H_2O_2/Lyocol > O_3. A colour removal of 85% and COD of more than 90% were already achieved after 10 min of reaction time for the photo-Fenton process. Therefore, the photo- Fenton process seems to be more appropriate as the pre-treatment method for decolorisation and detoxification of effluents from textile dyeing and finishing processes. Sulphate, nitrate, chloride, formate and oxalate were identified as main oxidation products.

Liu et al., (2010), evaluated the photocatalytic degradation of Reactive Brilliant Blue KN-R under UV irradiation in aqueous suspension of titanium dioxide under a variety of conditions. The degradation was studied by monitoring the change in dye concentration using UV spectroscopic technique. The decolorisation of the organic molecule followed a pseudo-first-order kinetics according to the Langmuir–Hinshelwood model. Under the optimum operation conditions, approximately 97.7% colour removal was achieved with significant reduction in TOC (57.6%) and COD (72.2%) within 3 hours. In aother study, Bergamini et al., (2009), investihated photocatalytic (TiO$_2$/UV) degradation of a simulated reactive dye bath (Black 5, Red 239, Yellow17, and auxiliary chemicals). After 30 min of irradiation, it was achieved 97% and 40% of colour removal with photocatalysis and photolysis, respectively. No mineralisation occurred within 30 min.

According to photocatalytic decolorisation studies, high rate of organic and color removal can be achieved. The main advantage of these processes is the usage of solar ligh. In another words, there is no energy need for hydroxyl rdical production. But, removal and recycling of semiconductors (TiO$_2$, ZnO etc.) from aqueous media is very important for both cost minimisation and effluent quality.

Finally, for true and good decision making on the treatment process, cost of all the compared processes should be calculated. In next step, cost evaluation of these processes are evaluated.

4.3 Cost evaluation of AOPs for colour removal

Cost evaluation is an important issue for decision making on a treatment process as much as process efficiency. Actual project costs can not be generalized; rather they are site-specific and thus must be developed for individual circumstances (Qasim et al., 1992). For a full-scale system, these costs strongly depend on the flow rate of the effluent and the configuration of the reactor as well as the nature of the effluent (Azbar et al., 2004). From these reasons, complete cost analysis of an AOP including treatment plant flow chart is too limited in literature. Azbar et al. 2005, Solmaz et al, Ustun et al, Yonar et al and Yonar, 2010 tried to explain the operational costs of examined AOPs. Average costs of applied processes are given in Table 2.

Process Type	Operational Cost (USD/m³)
Coagulation	0,07-0,20
Ozonation	4,21-5,35
Fenton process	0,23-0,59
Fenton-like process	0,48-0,57
Peroxane	5,02-5,85
UV/H_2O_2	1,26-4,56
UV/O_3	6,38-8,68
$UV/O_3/H_2O_2$	6,54-11,25

Azbar et al. 2005, Solmaz et al, Ustun et al, Yonar et al 2005 and Yonar, 2010

Table 2. Average operational costs of AOPs

In another study of Yonar 2010, treatment plant cost calculations were carried out according to Turkey conditions. The overall costs are represented by the sum of the capital costs, the operating costs and maintenance. For a full-scale system (200 m³/day, hand-printed textile wastewater), these costs strongly depend on the flow-rate of the effluent and the configuration of the reactor as well as the nature of the effluent (Azbar et al., 2004). Conventional treatment system (physical/ chemical/ biological treatment processes) and Fenton process (physical/Fenton processes) costs are summarized in this section for a meaningful explanation.

4.3.1 Capital costs

Capital costs of a treatment plant were calculated in four sub-stages: (1) constructional, (2) mechanical, (3) electrical, and (4) other costs.

Constructional costs of both treatment flow charts were computed by a Civil Engineering Office according to environmental design results. Constructional costs include excavation, reinforced concrete, buildings, excavator and crane rentals, electricity and labour costs. Land costs were excluded from the computations for the reason of industry own site usage. Mechanical and electrical costs are another chief and important capital costs for a treatment plant. Mechanical costs were determined by summing the costs of mechanical equipment purchase (coarse and fine screens, pumps, dosage pumps, mixers, chemical storage and handling tanks, blowers, diffusers, tank skimmers, filter-press etc.), pipes and fittings, material transportation and mechanical labour. Electrical costs contain automation, wiring, sensors (flow, pH, oxygen, ORP, level switches etc.) and electrical labour. Finally, other costs incorporate engineering design fee, charges and taxes, and profit and overhead. All equipment and material prices and labour costs were collected from different treatment plant equipment suppliers and engineering offices in Turkey.

ITEM	CAPITAL COSTS (Euro)	
	Conventional Treatment System	Fenton Process Treatment System
Construction Costs		
• Basin Constructions[1]	27 500 €	19 000 €
• Building Constructions[1,2]	13 000 €	13 000 €
• Rentals	6 300 €	4 700 €
• Electricity[3]	850 €	700 €
Mechanical Costs		
• Physical Unit Equipments[4]	10 800 €	10 800 €
• Chemical Unit Equipments[4]	12 800 €	23 500 €
• Biological Unit Equipments[4]	12 600 €	-
• Disinfection Unit Equipments[4]	700 €	-
• Sludge Unit Equipments[4]	15 500 €	15 500 €
• Piping Costs[4]	9 500 €	4 000 €
• Transportation and Rentals	2 500 €	2 000 €
Electrical Costs		
• Automation[5]	7 500 €	7 000 €
• Wiring[5]	1 500 €	1 500 €
• Sensors and Switches	3 600 €	2 200 €
SUB-TOTAL - A	124 650 €	103 900 €
Other Costs		
• Engineering Fee (5% of sub-total a)	6 233 €	5 195 €
• Charges	2 000 €	2 000 €
• Profit and overhead (15% sub-total a)	18 698 €	15 585 €
SUB-TOTAL - B	151 581 €	126 680 €
Taxes (VAT: 18% of sub-total - b)	27 285 €	22 803 €
TOTAL	**178 866 €**	**149 483 €**

[1]All construction costs include labour costs

[2]Buildings are designed (pre-fabric 200 m^2 closed area) as same capacity for both treatment plants including a small laboratory, chemical preparation and dosage units, blowers (for biological treatment unit) and sludge conditioning and filter-press units.

[3]1 kW = 0.087 Euro

[4]All mechanical costs include labour costs

[5]All electrical costs include labour costs

Table 3. Capital Cost Estimates of Conventional (Physical/Chemical/Biological) and Fenton Process (Physical/Fenton Process) Treatment Plants

Table 3 presents capital cost estimates for the conventional and Fenton process treatment plants designed on the basis of 200 m³/day. As shown in this table, the total capital cost estimates for conventional treatment plant and Fenton process treatment plant are 178 866 and 149 483 Euro, respectively. All equipment costs were provided including 2 years non-prorated warranty by all suppliers. But, sensors, switches and other spare parts were excluded from warranty. It can clearly be observed from the cost analysis that the specific costs for Fenton process treatment plant are about 16% lower than that of the conventional treatment plant alternative. On the other hand, constructional costs of the conventional treatment system are higher than Fenton process treatment alternative. But, mechanical and electrical capital cost trends can be regarded identical for both treatment alternatives. These cost differences originate from biological treatment unit, because activated sludge tank entails great construction area and more mechanical work effort.

4.3.2 Operation and maintenance costs
Operation and maintenance costs (O&M) include power requirement, chemicals, spare parts, wastewater discharge fees, plant maintenance and labour. Textile industry wastewater treatment plant sludges are accepted as a toxic and hazardous waste in Turkish Hazardous Wastes Control RegulationsAnonimous, 2005). Therefore, toxic and hazardous waste disposal costs and charges strongly depend on disposal technology and locations of the treatment plant and hazardous waste disposal plants. For these reasons, only sludge disposal costs were excluded from O&M cost estimations.

ITEM	Operating and Maintanence Costs (Euro/m³)	
	Conventional Treatment System	Fenton Process Treatment System
• Electrical power for processes and other facilities	0.24	0.17
• Spare part costs	0.04	0.03
• Chemicals	0.70	0.81
• Labour	0.34	0.34
• SUB-TOTAL	1.32	1.35
• Equipment repair, replacement and overhead (10% of sub-total)	0.13	0.135
TOTAL	1.452	1.485
REAGENT PRICES	Unit	Price (Euro)
• Hydrogen peroxide	Kg	0.55
• Sulphuric Acid	Kg	0.85
• Sodium Hydroxide	Kg	0.75
• Ferric Chloride	Kg	1.95
• Ferrous Sulphate	Kg	1.05
• Sodium Hypochlorite	Kg	0.20
• Polymer	Kg	2.35
• Electricity	kWh	0.087

Table 4. O&M costs of the studied treatment methods (cost of sludge disposal was excluded)

On the other hand, labour costs are very important part of O&M costs. Labour costs are facility-specific, and depend on the size, location and plant design. Therefore, labour costs may vary substantially (Pianta et al., 2000). These treatment plants can be considered as small treatment plants because of 200 m^3/day flow capacity. Accordingly, 9 working hours per day and a salary of 18 Euro/day (equal to minimum wage) for 2 workers and 32 Euro/day for operator and/or engineer were presumed, and the labour costs were calculated using a fixed rate of 0,34 Euro/m^3. Similar to labour costs, electricity and chemical prices are also country-specific. As shown in Table 3, the total costs of both conventional and Fenton process treatment plants were estimated as 1.452 Euro/m^3 and 1.485 Euro/m^3. According to these results, Fenton process treatment system O&M costs are slightly (3%) higher than conventional treatment system owing to relatively higher chemical usage of Fenton process treatment system. However, capital cost difference of both systems may afford operating cost difference for 15 years. The labour costs constitute about 23% of the overall O&M costs. On the other hand, electricity appears to be another important cost value for conventional system. Consequently, Fenton process has shown superior treatment and colour removal performances, and can be accepted as more economical choice for hand-printed textile wastewater treatment.

5. Conclusions

Advanced Oxidation Processes are promising alternative of traditional treatment proceeses for the treatment of textile effluent. Removal of colour and recalcitrant organic content of textile effluent can be achieved with the high efficiencies. Costs of AOPs are another point of view. In most cases, capital and operation and maintenance cost of AOPs are generally higher than traditional processes. But, Fenton process seems to be viable choice for textile wastewater treatment.

6. References

Akal Solmaz, S.K.; Birgul, A.; Ustun, G.E. & Yonar, T. (2006). Colour and COD removal from textile effluent by coagulation and advanced oxidation processes. *Color. Technol.*, Vol. 122, pp. 102–109, ISSN 1472-3581

Akal Solmaz, S.K.; Ustun, G.E.,Birgul,A. & Yonar, T., (2009). Advanced Oxidation Of Textile Dyeing Effluents: Comparison of Fe^{+2}/H_2O_2, Fe^{+3}/H_2O_2, O_3 and Chemical Coagulation Processes. *Fresenius Environmental Bulletin*,Vol.18 No.8, pp. 1424-1433, ISSN 1018-4619

Al-Kdasi, A.; Idris, A.; Saed, K. & Guan, C.T., (2004). Treatment of Textile Wastewater by Advanced Oxidation Processes – A Reviw. *Global Nest: the Int. J.* Vol.6, No.3, pp. 222-230, ISSN 1790-7632

Andreozzi, R.; Insola, A.; Caprio, V. & Amore, M.G.D., (1991). Ozonation of. Pyridine Aqueous Solution: Mechanistic and Kinetic Aspects. *Water res*, Vol.25, No.2, pp. 655-659, ISSN 0043-1354

Andrezozzi, R.; Caprio, V.; Insola, A. & Marotta, R., (1999). Advanced Oxidation Processes (AOP) for Water Purifiction and Recovery. *Catalysis Today*, Vol.53, pp. 51-59, ISSN 0920-5861

Anonymous (2005) Turkish Hazardous Wastes Control Regulation, Ministry of Environment and Forestry, Turkish Republic.

APHA-AWWA-WEF, (1995). Standard Method for the Examination of Water and Wastewater, 19th ed. American Public Health Association, Washington DC.

Arslan, I. & Balcioglu, I.A., (2000). Effect of common reactive dye auxiliaries on the ozonation of dyehouse effluents containing vinylsulphone and aminochlorotriazine dyes. *Desalination*, Vol.130, pp. 61-71, ISSN 0011-9164

Arslan, I. (2000). Treatment of Reactive Dye-Bath Effluents by Heterogeneous and Homogeneous Advanced Oxidation Processes, Submitted to the Institute of Environmental Sciences in partial fulfillment of the requirements for the degree of Doctor of Philosophy in Environmental Technology, Bogazici University.

Amini, M.; Arami, M.; Mahmoodi, N.M. & Akbari, A. (2011). Dye removal from colored textile wastewater using acrylic grafted nanomembrane. *Desalination*. Vol.267, pp. 107–113, ISSN 0011-9164

Atchariyawut, S.; Phattaranawik, J.; Leiknes, T. & Jiraratananon, R., (2009). Application of ozonation membrane contacting system for dye wastewater treatment. *Sep. Purif. Technol.* Vol.66, pp. 153–158 , ISSN 1383-5866

Azbar, N.; Yonar, T. and Kestioglu, K., (2004). Comparison of various advanced oxidation processes and chemical treatment methods for COD and colourremoval from a polyester and acetate fiber dyeing effluent. *Chemosphere* Vol.55, pp. 35–43, ISSN 0045-6535

Azbar, N.; Kestioğlu, K. & Yonar, T., (2005). Application of Advanced Oxidation Processes (AOPs) to Wastewater Treatment. Case Studies: Decolourization of Textile Effluents, Detoxification of Olive Mill Effluent, Treatment of Domestic Wastewater, *Ed. A.R. BURK, Water Pollution: New Research,* pp. 99-118, New York, Nova Science Publishers, ISBN-1-59454-393-3

Bautista, P.; Mohedano, A.F.; Gilarranz, M.A.; Casas, J.A. & Rodriguez, J.J. (2007). Application of Fenton oxidation to cosmetic wastewaters treatment. *J.Hazard.Mater.* Vol.143, pp. 128-134, ISSN 0304-3894

Baxendale, J.H. & Wilson, J.A., (1956). The photolysis of hydrogen peroxide at high light intensities. *Trans. Faraday Soc.* Vol.53, pp. 344-356, ISSN 0956-5000.

Benitez, F. J.; Acero, J. L.; Real, F. J.; Rubio, F. J. & Leal, A.I. (2001). The role of hydroxyl radicals for the decomposition of p-hydroxy phenylacetic acid in aqueous solutions. *Water Res.* Vol.35, pp. 1338-1343, ISSN 0043-1354

Benitez, J.F.; Beltran-Henadian, J.; Acero, J.L. & Rubio, F.J., (2000). Contribution of Free Radicals To Chlorophenols Decomposition by Several Advanced Oxidation Process. *Chemosphere*, Vol.41, pp. 1271-1277, ISSN 0045-6535

Bergaminia, R.B.M.; Azevedob, E.B. & Raddi de Araújo, R.D., (2009). Heterogeneous photocatalytic degradation of reactive dyes in aqueous TiO_2 suspensions: Decolorization kinetics. *Chemical Engineering Journal*. Vol.149, pp. 215–220, ISSN 1385-8947

Bes-Piá, A.; Mendoza-Roca, J.A.; Roig-Alcover, L.; Iborra-Clar, A.; Iborra-Clar, M.I. & Alcaina- Miranda, M.I., (2003). Comparison between nonofiltration and ozonation of biologically treated textile wastewater for its reuse in the industry. *Desalination*, 157, 81-86, ISSN 0011-9164

Bhattacharjee, S. & Shah, Y.T., (1998). Mecanism for Advance Photooxidation of Aqueous Organic Waste Compounds. *Rev. Chem. Eng.* Vol. 14, pp. 1-8, ISSN 0167-8299

Bigda, R.J. (1995). Consider Fenton's chemistry for wastewater treatment. *Chem.Eng.Prog.* Vol.91, pp. 62-66, ISSN 0360-275

Bossmann, S.H.; Oliveros, E.; Göb, S.; Kantor, M.; Göppert, A.; Lei, L.; Yue, P.L. & Braun, A.M., (2001). Degradation of polyvinyl Alchol "PVA" by Homogeneous and Heterogeneous Photocatalysis Applied to the Photochemical Enhanced Fenton Reaction. *Wat. Sci. Tech.*, Vol.44, pp. 257-262, ISSN 0273-1223

Bolton, J. R., (2001). Ultraviolet Applications Handbook. 2nd Ed., Bolton Photosciences Inc., 628 Cheriton Cres., NW, Edmonton, AB, Canada T6R 2M5.

Braun, A.M.; Jacob, L. & Oliveros, E., (1993). Advanced oxidation processes — concepts of reactor design. J. *Water SRT Aqua*, Vol. 42, No.3, pp. 166–173, ISSN 0003-7214

Bulut, Y. & Aydin, H., (2006). A kinetics and thermodynamics study of methylene blue adsorption on wheat shells. *Desalination*. Vol.194, pp. 259–267, ISSN 0011-9164

Cisneros, R.L.; Espinoza, A.G. & Litter, M.I., (2002). Photodegradation of an azo dye of the textile industry. *Chemosphere*, Vol. 48, pp. 393–399, ISSN 0045-6535

Contreras, S.; Rodriguez, M.; Chamarro, E.; Esplugas, S. & Casado,J. (2001), Oxidation of Nitrobenzene by UV/O₃: the Influence of H₂O₂ and Fe(III) Experiences in a Pilot Plant. *Wat. Sci. Tech.*, Vol. 44, pp. 39-46, ISSN 0273-1223

Corriera. V.M.; Stephenson. T. & Judd, S.J. (1994). Characterisation of textile wastewaters-A review. *Environ Technol.* Vol.15, pp. 91.7 -919. ISSN 0959-3330

Demirbas, E.; Kobya, M.; Oncel, S. & Sencan, S., (2002). Removal of Ni(II) from aqueous solution by adsorption onto hazelnut shell activated carbon: equilibrium studies. *Bioresource Technol.* Vol.84, pp. 291–293, ISSN 0960-8524

El-Dein, A.M.; Libra, J.A. & Wiesmann, U., (2003). Mechanism and kinetic model for the decolorization of the azo dye reactive black 5 by hydrogen peroxide. *Chemosphere*. Vol.52, pp. 1069–1077, ISSN 0045-6535

EPA, (1998). *Handbook on Advanced Photochemical Oxidation Process,*, US. EPA, Washington, DC.

EPA, (2001). *Handbook on Advanced Non-Photochemical Oxidation Process*, US. EPA, Washington, DC.

Erswell. A., Brouckaert. C.J. & Buckley, C.A. (2002) The reuse of reactive dye liquors using charged ultrafiltration membrane technology. *Desalination*. Vol.143, pp. 243-253, ISSN 0011-9164

Fang, S.; Jiang, Y.; Wang, A.; Yang, Z. & Li, F., (2004). Photocatalytic performance of natural zeolite modified by TiO₂, *China. Non-Metallic Mines*. Vol. 27, No.1, pp. 14- 21, ISSN 1007-9386

Fenton, H.J.H. (1894). Oxidation of tartaric acid in presence of iron. *J. Chem. Soc., Trans.* Vol.65, No.65, pp. 899–911, ISSN 0956-5000

Forni, L.; Bahnemann, D. & Hart, E.J., (1982). Mechanism of the Hydroxide. Ion Initiated Decomposition of Ozone in Aqueous Solution. *J. Phys. Chem.*, Vol.86, pp. 255-259, ISSN 0022-3654

Galindo, C. & Kalt, A., (1998). UV-H2O2 oxidation of monoazo dyes in aqueous media: a kinetic study. *Dyes Pigments*. Vol. 40, pp. 27-35, ISSN 0143-7208

Galindo, C. ; Jacques, P. & Kalt, A., (2001). Photochemical and photocatalytic degradation of an indigolid dye: A case study of acid blue 74 (AB74). *Journal of Photochemistry and Photobiology A: Chemistry*. Vol. 141, pp. 47-56, ISSN 1010-6030

Georgiou, D.; Melidis, P.; Aivasidis, A. & Gimouhopoulos, K., (2002). Degradation of Azo-Reactive Dyes by Ultraviolet Radiation in The Presence of Hydrogen Peroxide. *Dyes and Pigments*.Vol.52, pp. 69–78, ISSN 0143-7208

Glaze, W.H. & KANG, J.W., (1989). Advanced Oxidation Processes: Test of a Kinetic Model for the Oxidation of Organic Compounds with Ozone and Hydrogen Peroxide in a Semi-batch Reactor. *Ind. Eng. Chem. Res.,* Vol.28; No.11, pp. 1580-1587, ISSN 0888-5885

Gogate, P.R. & Pandit, A.B., (2004a). A Review of Imperative Technologies for Wastewater Treatment I: Oxidation Technologies at Ambient Cconditions. *Advances in Environmental Research*. Vol.8, pp. 501-551, ISSN 1093-0191

Gogate, P.R. & Pandit, A.B., (2004b). A Review of Imperative Technologies for Wastewater Treatment II: Hybrid Methods. *Advances in Environmental Research*. Vol.8, pp. 553-597, ISSN 1093-0191

Goi, A. & Trapido, M., (2002). Hydrogen peroxide photolysis, Fenton reagent and photo-Fenton for the destraction of nitrophenol: a comparative study. *Chemosphere*. Vol.46, pp. 913–922, ISSN 0045-6535

Gonzaleza, M.G.; Oliveros, E.; Worner, M. & Braun, A.M., (2004). Vacuum-ultraviolet photolysis of aqueous reaction systems. *Journal of Photochemistry and Photobiology C: Photochemistry Reviews*, Vol.5, pp. 225–246, ISSN 1389-5567

Grau, P. (1991) Textile industry wastewaters treatment. *Water Sci. Technol.* Vol. 24, pp. 97 - 103, ISSN 0273-1223

Kestioglu, K.; Yonar, T. & Azbar, N., (2005). Feasibility of Physico-Chemical Treatment and Advanced Oxidation Processes (AOPs) as a Means of Pretreatment of Olive Mill Effluent (OME). *Process Biochemistry*, Vol.40, pp. 2409-2416, ISSN 1359-5113

He, F. & Lei, L. Degradation kinetics and mechanisms of phenol in Photo-Fenton Process, *Journal of Zhejiang University Science (J Zhejiang Univ SCI)*, Vol.5, No.2, pp. 198-205, ISSN 1862-1775

Hoigne, J. & Bader, H. (1983). Rate constants of reaction of ozone with organic and inorganic compounds in water. Part II. Dissociating organic compounds. *Water Res.*, Vol. 17, pp. 185-194, ISSN 0043-1354

Hoigne, J. (1998). Chemistry of aqueous ozone and transformation of pollutants by ozone and advanced oxidation processes, in: J. Hrubec (Ed.), *The Handbook of Environmental Chemistry*, Vol. 5, Part C, Quality and Treatment of Drinking Water, Part II, Springer, Berlin Heidelberg

Horsch, F., (2000). Oxidation Eines Industriellen Mischabwassers mit Ozon und UV/H2O2. *Vom Wasser*, Vol.95, pp. 119-130, ISSN

Hsiao, C.Y.; Lee, C.L. & Ollis, D.F., (1983), Heterogeneous photocatalysis: degradation of dillute solutions of dichlorometane (CH_2Cl_2), chloroform ($CHCl_3$) and carbon tetrachloride (CCl₄) with illuminated TiO_2 photocatalyst, *J. Catal.* Vol.82, pp. 418-423, ISSN 0021-9517

Huang, Y.; Tsai, S.; Huang, Y. & Chen, C. (2007). Degradation of commercial azo dye reactive Black B in photo/ferrioxalate system, *Journal of Hazardous Materials.* Vol.140, pp. 382–388, ISSN 0304- 3894

Hung-Yee, S. & Ching-Rong, H., (1995). Degradation of commercial azo dyes in water using ozonation and UV enhanced ozonation process. *Chemosphere*, Vol.31, pp. 3813- 3825, ISSN 0045-6535

Ince, N.H., (1999). Critical effect of hydrogen peroxide in photochemical dye degradation. *Water Res.* Vol. 33, pp. 1080–1084, ISSN 0043-1354

Kang, Y.W. &Hwang, K.Y., (2000). Effect of reaction conditions on. the oxidation efficiency in the Fenton process. *Water Res.* Vol. 34, pp. 2786-2790, ISSN 0043-1354

Kang, S., Liao, C. & Po, S., (2000). Decolorization of textile wastewater by photo-fenton oxidation technology. *Chemosphere*, Vol. 41, pp. 1287-1294, ISSN 0045-6535

Kestioglu, K., Yonar, T. & AZBAR, N., (2005). Feasibility of Physico-Chemical Treatment and Advanced Oxidation Processes (AOPs) as a Means of Pretreatment of Olive Mill Effluent (OME). *Process Biochemistry*, Vol.40, pp. 2409-2416, ISSN 0032-9592

Kormann, C.; Bahnemann, D.F. & Hoffmann, M.R., (1991), Photolysis of chloroform and other organic molecules in aqueous TiO_2 suspensions. *Environ. Sci. Technol.* Vol.25, pp. 494-500, ISSN 0013-936X

Kuo, W.G., (1992). Decolorizing Dye Wastewater with Fenton's Reagent. *Wat. Res.* Vol.26, pp. 881-886, ISSN 0043-1354

Kwon, B.G.; Lee, D.S.; Kang, N. & Yoon, J., (1999). Characteristics of *p*-chlorophenol oxidation by Fenton's reagent, *Water Res.* Vol.33, pp. 2110-2118, ISSN 0043-1354

Lee, H. & Shoda, M., (2008). Removal of COD and color from livestock wastewater by the fenton method. *J. Hazard. Mater.* Vol.153, pp. 1314-1319, ISSN 0304-3894

Li, F.; Sun, S.; Jiang, Y.; Xia, M.; Sun, M. & Xu, B. (2008). Photodegradation of an azo dye using immobilized nanoparticles of TiO2 supported by natural porous mineral. *J. Hazard. Mater.* Vol. 152, pp. 1037–1044, ISSN 0304-3894

Lin, S.H. & Lo, C.C., (1997). Fenton process for treatment ofdesizing wastewater. *Water Res.* Vol. 31, pp.2050-2058, ISSN 0043-1354

Liao, C.H.; Kang, S.F. & Hung, H.P., (1999). Simultaneous removals of COD and color from dye manufacturing process wastewater using photo-fenton oxidation process. *J. Environ. Sci. Health A* Vol.34 , pp. 989-1010, ISSN 1532-4117

Liou, M.J.; Lu, M.C. & Chen, J.N., (2004). Oxidation of TNT by photo-Fenton process. *Chemosphere.* Vol.57, pp.1107–1114, ISSN 0045-6535

Liu, R.; Chiu, H.M.; Shiau, C.; Yeh, R.Y. & Hung, Y., (2007). Degradation and sludge production of textile dyes by Fenton and photo-Fenton processes. *DyesPigments.* Vol. 73, pp. 1-6, ISSN 0143-7208

Liu, Y.; Hua, L. & Li, S., (2010). Photocatalytic degradation of Reactive Brilliant Blue KN-R by TiO2/UV process. *Desalination.* Vol. 258, pp. 48–53, ISSN 0011-9164

Little, L.W.; Lamb, J.C.; Chillingworth, M.A. and Durkin, W.B., (1974). Acute toxicity of selected commercial dyes to the fathead minnow and evaluation of biological treatment for reduction of toxicity. *In: Proc. 29th Ind. Waste Conf.,* Purdue University, Lafayette, IN, USA, pp. 524–534.

Mahmoodi, N.M. & Arami, M., (2006). Bulk phase degradation of acid Red 14 by nanophotocatalysis using immobilized titanium (IV) oxide nanoparticles. *J. Photochem. Photobiol. A: Chem.* Vol. 182, pp. 60–66, ISSN 1010-6030

Mahmoodi, N.M. & Arami, M., (2008). Modeling and sensitivity analysis of dyes adsorption onto natural adsorbent from colored textile wastewater. *J. Appl. Polym. Sci.* Vol.109, pp. 4043–4048, ISSN 0021-8995

Mahmoodi, N.M. & Arami, M., (2009). Numerical finite volume modeling of dye decolorisation using immobilized titania nanophotocatalysis. *Chem. Eng. J.* Vol.146, 189–193, ISSN 1385-8947

Mahmoodi, N.M. & Arami, M., (2009). Degradation and toxicity reduction of textile wastewater using immobilized titania nanophotocatalysis. *J. Photochem. Photobiol. B: Biol.* Vol.94, pp. 20–24, ISSN 1011-1344

Mahmoodi, N.M. & Arami, M., (2010). Immobilized titania nanophotocatalysis: degradation, modeling and toxicity reduction of agricultural pollutants. *J. Alloy. Compd.* Vol.506, pp. 155–159, ISSN 0925-8388

Mandal, A.; Ojha, K.; De, A.K. & Bhattacharjee, S., (2004). Removal of catechol from aqueous solution by advanced photo-oxidation process. *Chemical Engineering Journal,* Vol.102, pp. 203–208, ISSN 1385-8947

Metcalf, Eddy, Inc. (2003), Wastewater engineering treatment and reuse, Fourth Edition, *McGraw-Hill,* New York ISBN 10: 007041677X

Mokrini, M.; Oussi, D. & Esplugas, S., Oxidation of Aromatic Compounds with Uv Radiation/Ozone/Hydrogen Peroxide. *Water Sci. Ttechnol.,* Vol.35, pp. 95-102, ISSN 0273-1223

Mozia, S.; Morawski, A.W.; Toyoda, M. & Inagaki, M., (2008). Effectiveness of photodecomposition of an azo dye on a novel anatase-phase TiO_2 and two commercial photocatalysts in a photocatalytic membrane reactor (PMR). *Sep. Purif. Technol.* 63, pp. 386–391, ISSN 1383-5866

Muruganandham, M. & Swaminathan, M., (2004). Decolourisation of reactive Orange 4 by Fenton and photo-Fenton oxidation technology. *Dyes Pigments* Vol.63, pp. 315–321, ISSN 0143-7208

Muthukumar, M.; Selvakumar, N. & Venkata, R.J., (2001). Effect of dye structure on decoloration of anionic dyes by using ozone. *International Ozone Association, Proceedings of the 15th World Congress,* London, 2001.

Neamtu, M.; Siminiceanu, I.; Yediber, A. & Kettrup, A. (2002). Kinetics of decolorization and mineralization of reactive azo dyes in aqueous solution by the UV/H_2O_2 oxidation. *Dyes Pigm.* Vol. 53, pp. 93–99, ISSN 0143-7208

Neamtu, M.; Yediler, A.; Siminiceanu, I.; Macoveanu, M. & Kettrup, A., (2004). Decolorization of Disperse Red 354 Azo Dye in Water by Several Oxidation Processes-A Comparative Study. *Dyes and Pigments,* Vol. 60, pp. 61-68, ISSN 0143-7208

Nesheiwat, F.K. & Swanson, A.G., (2000). Clean contaminated sites using. Fenton's reagent. *Chem.Eng. Prog.* Vol. 96, No.4, pp. 61-66, ISSN 0360-7275

Nigam, P.; Armour, G.; Banat, I.M.; Singh, D. & Marchant, R., (2000). Physical removal of textile dyes and solid state fermentation of dye-adsorbed agricultural residues. *Bioresour. Technol.* 72, 219–226, ISSN 0960 8524

O'Neill, C.; Freda, R.H.; Dennis, L.H.; Lourenco Nidia, D; Helena, M.P.; & Delee, W., (1999). Colour in Textile Effluents-Sources, Measurement, Discharge Consents and Simulation: A Review, *Journal of Chemical Technology and Biotechnology*, Vol. 74, pp. 1009-1018, ISSN 1097-4660

Pagga, U. & Brown, D. (1986). The degradation of dyestuffs: Part II Behaviour of dyestuffs in aerobic biodegradation tests. *Chemosphere.* Vol.15, No.4, pp. 479-491, ISSN 0045-6535

Perkowski J. & Kos L., (2003). Decolouration of model dye house wastewater with advanced oxidation processes. *Fibres and Textiles in Eastern Europe*, Vol. 11, pp. 67-71. ISSN 1230-3666

Pianta, R.; Boller, M.; Urfer, D.; Chappaz, A. & Gmünder, A. (2000). Costs of conventional vs. membrane treatment for karstic spring water. *Desalination.* Vol.131, pp. 245-255, ISSN 0011-9164

Pignatello, J.J., (1992). Dark and photassisted Fe^{+3}-catalyzed degradation of chlorophenoxy herbicides by hydrogen peroxide. *Environ.Sci.Technol.* Vol.26, pp. 944-951, ISSN 0013-936X

Qasim, S.R.; Lim, S.W.D.; Motley, E.M. & Heung, K.G., (1992). Estimating costs for treatment plant construction. *Journal American Water Works Association.* Vol.84, pp. 56-62, ISSN 1551-8833

Rein M., (2001). Advanced oxidation processes – current status and prospects. *Proc. Estonian Acad, Science Chemistry*, Vol.50, pp. 59–80, ISSN 1406-0124

Robinson, T.; McMullan, G.; Marchant, R. & Nigam, P., (2004). Remediation of dyes in textile effluent: a critical review on current treatment technologies with a proposed alternative. *Bioresource Technol.* Vol.77, pp. 247-255, ISSN 0960-8524

Sedlak, D.L. & Andren, A.W., (1991). Oxidation of chlorobenzene with Fentons reagent. *Environ.Sci.Technol.* Vol. 25, pp. 777-782, ISSN 0013-936X

Sevimli MF & Sarikaya, H.Z., (2002). Ozone treatment of textile effluents and dyes: effect of applied ozone dose, pH and dye concentration. *J Chem Technol Biotechnol* Vol .77, No.7, pp. 842-850, ISSN 0268-2575

Selcuk, H. (2005). Decolorisation and detoxification of textile wastewater by ozonation and coagulation processes. *Dyes and Pigments*, Vol.64, pp. 217-222, ISSN 0143-7208

Shu, H.Y.; Huang, C.R. & Chang, M.C., (1994). Decolorization of mono-azo dyes in wastewater by advanced oxidation process: a case study of acid red 1 and acid yellow 23. *Chemosphere.* Vol.29, pp. 2597–2607, ISSN 0045-6535

Shu, H.Y. & Chang, M.C., (2005). Decolorisation effects of six azo dyes by O3, UV/O3 and UV/H2O2 processes. *Dyes and Pigments.* Vol.65, pp. 25-31, ISSN 0304-3894

Slokar, Y.M. & Majcen Le Marechal, A. (1998). Methods of decoloration of textile wastewaters. *Dyes Pigments.* Vol.37, pp. 335–356, ISSN 0143-7208

Somensia, C.A.;. Simionattoa, E.L.; Bertoli, S.L.; Wisniewski Jr.A. & Claudemir M. Radetski, (2010). Use of ozone in a pilot-scale plant for textile wastewater pre-treatment: Physico-chemical efficiency, degradation by-products identification and environmental toxicity of treated wastewater. *Journal of Hazardous Materials*, Vol.175, pp. 235–240, ISSN 0304-3894

Steahlin, J. & Hoigne, J., (1982). Decomposition of Ozone in Water: Rate of Initiation by Hydroxide Ions and Hydrogen Peroxide. *Environ. Sci. Technol.*, Vol.16, pp. 676-681, ISSN 0013-936X

Staehelin, J. & Hoigne, J. (1985). Decomposition of ozone in water in the presence of organic solutes acting as promoters and inhibitors of radical chain reactions. *Env. Sci. Technol.*, Vol. 19, pp. 1206-1213, ISSN 0013-936X

Stowell, J.P. & Jensen, J.N., (1991). Dechlorination of Chlorendic Acid with Ozone. *Water Res.*,Vol.25, pp. 83-90, ISSN 0043-1354

Sun, Y. & Pignatello, J.J., (1993). Photochemical reactions involved in the total mineralisation o 2-4-D by $Fe^{+3}/H_2O_2/UV$. *Environ.Sci. Technol.*, Vol.27, No.2, pp. 304-310, ISSN 0013-936X

Tang, W.Z. & Huang, C.P., (1996). 2,4-dichlorophenol oxidation kinetics by Fenton's reagent. Environ. Technol. Vol.17, pp. 1371-1378, ISSN 0959-3330

Techcommentary (1996), Advanced oxidation processes for treatment of industrial wastewater. An EPRI community environmental centre publ. No. 1.

Torrades, F.; Perez, M.; Mansilla, H.D. & Peral, J., (2003). Evperimental Design of Fenton and photo Fenton fort he Treatment of Cellulose Bleaching Effluents *Chemosphere.* Vol.53, 1211–1220, ISSN 0045-6535

Uygur, A. & Kok, E. (1999). Decolorisation treatments of azo dye waste waters including dichlorotriazinyl reactive groups by using advanced oxidation method. *J.S.D.C.*, Vol.115, pp. 350-354, ISSN 1478-4408

Venceslau, M.C.; Tom, S.; & Simon, J.J., (1994). Characterisation of textile wastewater-a review, *Environ. Technol.*, Vol.15, pp. 917-929, ISSN 0959-3330

Vendevivere, P.C.; Bianchi, R. & Verstraete, W., (1998). Treatment and reuse of wastewater from the textile wet-processing industry: reviewof emerging technologies. *J. Chem. Technol. Biotechnol.*, Vol.72, pp. 289-302, ISSN0268-2575

Venkatadri, R. & Peters, R.V., (1993). Chemical oxidation technologies: ultraviolet light/hydrogen peroxide, Fenton's reagent and titanium dioxide-assisted photocatalysis. *Hazard. Waste Hazard.Mater* Vol.10, pp. 107-149, ISSN 0882-5696

Willmott, N.J. (1997). The use of bacteria-polymer composites for the removal of colour from reactive dye effluents. *PhD thesis, UK, Univeristy of Leeds.*

Willmott, N.J.; Guthrie, J. & Nelson, G. (1998). The Biotechnology Approach to Colour Removal from Textile Effluent. *J. of the Soc. of Dyers and Colourists*, Vol.114, pp. 38-41, ISSN 1478-4408

Yang, W.; Wu, D. & Fu, R., (2008). Effect of surface chemistry on the adsorption of basic dyes on carbon aerogels. *Colloid Surf. A: Physicochem. Eng. Aspects.* Vol. 312, pp. 118–124, ISSN 0927-7757

Yonar, T.; Yonar, G.K.; Kestioglu, K. & Azbar, N., (2005). Decolorisation of Textile Effluent Using Homogeneous Photochemical Oxidation Processes. *Colour. Technol.* Vol.121, pp. 258-264, ISSN 1472-3581

Yonar, T., (2010). Treatability Studies on Traditional Hand-Printed Textile Industry Wastewaters Using Fenton and Fenton-Like Processes: Plant Design and Cost Analysis. *Fresenius Environmental Bulletin,* Vol.19, No.12 2758-2768, ISSN 1018-4619

Zahraa, O.; Chen, H.Y. & Bouchy, M., (1999). Adsorption and photocatalytic degradation of 1,2-dichloroethane on suspended TiO_2. *J. Adv. Oxid. Technol.,* Vol.4, pp. 167-173, ISSN 1203-8407

2

Textile Dyeing Wastewater Treatment

Zongping Wang, Miaomiao Xue, Kai Huang and Zizheng Liu
Huazhong University of Science and Technology
China

1. Introduction

Textile industry can be classified into three categories viz., cotton, woolen, and synthetic fibers depending upon the used raw materials. The cotton textile industry is one of the oldest industries in China.

The textile dyeing industry consumes large quantities of water and produces large volumes of wastewater from different steps in the dyeing and finishing processes. Wastewater from printing and dyeing units is often rich in color, containing residues of reactive dyes and chemicals, such as complex components, many aerosols, high chroma, high COD and BOD concentration as well as much more hard-degradation materials. The toxic effects of dyestuffs and other organic compounds, as well as acidic and alkaline contaminants, from industrial establishments on the general public are widely accepted. At present, the dyes are mainly aromatic and heterocyclic compounds, with color-display groups and polar groups. The structure is more complicated and stable, resulting in greater difficulty to degrade the printing and dyeing wastewater (Shaolan Ding et al.,2010).

According to recent statistics, China's annual sewage has already reached 390 million tons, including 51% of industrial sewage, and it has been increasing with the rate of 1% every year. Each year about 70 billion tons of wastewater from textile and dyeing industry are produced and requires proper treatment before being released into the environment (State Environmental Protection Administration ,1994).

Therefore, understanding and developing effective printing-dye industrial wastewater treatment technology is environmentally important.

1.1 Textile printing and dyeing process

Textile Printing and dyeing processes include pretreatment, dyeing / printing, finishing and other technologies.

Pre-treatment includes desizing, scouring, washing, and other processes. Dyeing mainly aims at dissolving the dye in water, which will be transferred to the fabric to produce colored fabric under certain conditions. Printing is a branch of dyeing which generally is defined as 'localized dyeing' i.e. dyeing that is confirmed to a certain portion of the fabric that constitutes the design. It is really a form of dyeing in which the essential reactions involved are the same as those in dyeing. In dyeing, color is applied in the form of solutions, whereas color is applied in the form of a thick paste of the dye in printing. Both natural and synthetic textiles are subjected to a variety of finishing processes. This is done to improve specific properties in the finished fabric and involves the use of a large number of finishing agents for softening, cross-linking, and waterproofing. All of the finishing processes contribute to water pollution. In addition, in different circumstances, the singeing, mercerized, base reduction, and other processes should have been done before dyeing/printing.

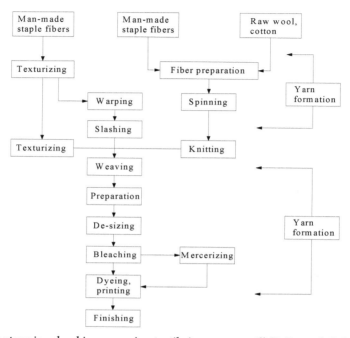

Fig. 1. Various steps involved in processing textile in a cotton mill (B. Ramesh Babu et al.,2007).

In the textile dyeing industry, bleaching is an important process. It has three technologies: sodium hypochlorite bleaching; hydrogen peroxide bleaching and sodium chlorite bleaching. Sodium hypochlorite bleaching and sodium chlorite bleaching are the most commonly used processes. Normal concentration of chlorine dioxide in bleaching effluent is 10-200 mg/L. As chlorine dioxide is a strong oxidant, it is very corrosive and toxic as well. The typical printing and dyeing process is shown in Fig. 1 and the main used fiber dyes at present have been shown in Table 1 (Kelu Yan,2005).

The variety of fiber	The commonly used dyes
Cellulose fiber	Direct dyes, Reactive dyes,Vat dyes, Sulfide dyes, Azo dyes
Wool	Acid dyes
Silk	Direct dyes,Acid dyes
Polyester	Azo dyes,Disperse dyes
Polyester-cotton	Disperse / Vat dyes,Disperse / Insoluble dye
polyacrylonitrile fiber	Cationic dyes, Disperse dyes
polyacrylonitrile fiber -wool	Cationic dyes,Acid dyes
vinylon	Direct dyes,Vat dyes,Sulfur dyes,Acid dyes,

Table 1. The varieties of common used fiber

1.2 Production of textile industry pollution

Textile Printing and dyeing processes include pre-treatment, dyeing and printing, finishing. The main pollutants are organic matters which come from the pre-treatment process of pulp, cotton gum, cellulose, hemicellulose and alkali, as well as additives and dyes using in dyeing and printing processes. Pre-treatment wastewater accounts for about 45% of the total, and dyeing/printing process wastewater accounts for about 50%~55%, while finishing process produces little.

In China, chemical fiber accounts for about 69% of total in which polyester fibers accounts for more than 80%. Cotton accounts for 80% of the natural fiber production. Therefore, the dyeing wastewater analysis of production and pollution is based on these two fibers.

Pre-treatment of cotton includes desizing and scouring. The main pollutants are the impurities in the cotton, cotton gum, hemicellulose and the slurry, alkali in weaving process. The current average COD concentration in the pre-treatment is 3000 mg/L. The main pollutants in dyeing/printing are auxiliaries and the residual dyes. The average concentration of COD is 1000 mg/L and the total average concentration is 2000 mg/L after mixing.

Pre-treatment of polyester fibers mainly involves in the reduction with alkali. The so-called reduction is treating the polyester fabric with 8% of sodium hydroxide at 90 $^\circ$C for about 45 minutes. Some polyester fabrics will peel off and decompose into terephthalic acid and ethylene glycol so that a thin polyester fabric will have the feel of silk. This process can be divided into continuous and batch type. Taking the batch type as an example, the concentration of COD is up to 20000 mg/L-60000 mg/L. The wastewater from reduction process may account for only 5% of the volume of wastewater, while COD accounts for 60% or more in the conventional dyeing and finishing business.

The chroma is one pollutant of the wastewater which causes a lot of concerns. In the dyeing process, the average dyeing rate is more than 90%. It means that the residual dyeing rate in finishing wastewater is about 10%, which is the main reason of contamination. According to the different dyes and process, the chroma is 200 to 500 times higher than before.

pH is another factor of the dyeing wastewater. Before the printing and the dyeing process, pH is another factor ,the pH of dyeing wastewater remains between 10 to 11 when treated by alkali at high temperature around 90°C in the process of desizing, scouring and mercerization. Polyester base reduction process mainly uses sodium hydroxide, and the total pH is also 10 to 11. Therefore, most dyeing water is alkaline and the first process is to adjust the pH value of the textile dyeing wastewater.

The total nitrogen and ammonia nitrogen come from dyes and raw materials, which is not very high, about 10 mg/L. But the urea is needed while using batik techniques. Its total nitrogen is 300 mg/L, which is hard to treat. The phosphorus in the wastewater comes from the phosphor detergents. Considering the serious eutrophication of surface water, it needs to be controlled. Some enterprises use trisodium phosphate so that the concentration of phosphorus will reach 10 mg/L. So, this phosphorus must be removed in the pre-treatment.

In the production process, suspended substance comes from fiber scrap and undissolved raw materials. It will be removed through the grille, grid, etc. The suspended solids (SS) in the outflow mainly come from the secondary sedimentation tank, whose sludge has not been separated completely which will reach 10-100 mg/L as usual.

Sulfide mainly comes from the sulfur, which is a kind of cheap and qualified dye. Due to its toxicity, it has been forbidden in developed countries. However, in China, some enterprises are still using it, so it has been included in the wastewater standards. The sulfide in the wastewater is about 10 mg/L. There are two main sources of hexavalent chromium. Cylinder engraving makes the wastewater containing hexavalent chromium. However, this technology has not been used. Another possible source is the use of potassium dichromate additive in hair dyeing process. Aniline mainly comes from the dyes. The color of the dye comes from the chromophore. Some dyes have a benzene ring, amino, etc., which will be decomposed in the wastewater treatment process.

The potential specific pollutants from textile Printing and dyeing is shown in Table 2 (C. All`egre et al.,2006).

Process	Compounds
Desizing	Sizes, enzymes, starch, waxes, ammonia.
Scouring	Disinfectants and insecticides residues, NaOH, surfactants, soaps, fats, waxes, pectin, oils, sizes, anti-static agents, spent solvents, enzymes.
Bleaching	H_2O_2, AOX, sodium silicate or organic stabiliser, high pH.
Mercerizing	High pH, NaOH
Dyeing	Colour, metals, salts, surfactants, organic processing assistants, sulphide, acidity/alkalinity, formaldehyde.
Printing	Urea, solvents, colour, metals.
Finishing	Resins, waxes, chlorinated compounds, acetate, stearate, spent solvents, softeners.

Table 2. Specific pollutants from textile and dyeing processing operations

1.3 Textile dyeing wastewater risk

Discharged wastewater by some industries under uncontrolled and unsuitable conditions is causing significant environmental problems. The importance of the pollution control and treatment is undoubtedly the key factor in the human future. If a textile mill discharges the wastewater into the local environment without any treatment, it will has a serious impact on natural water bodies and land in the surrounding area. High values of COD and BOD_5, presence of particulate matter and sediments, and oil and grease in the effluent causes depletion of dissolved oxygen, which has an adverse effect on the aquatic ecological system.

Effluent from textile mills also contains chromium, which has a cumulative effect, and higher possibilities for entering into the food chain. Due to usage of dyes and chemicals, effluents are dark in color, which increases the turbidity of water body. This in turn hampers the photosynthesis process, causing alteration in the habitat (Joseph Egli,2007).

1.4 The textile industry standards for water pollutants

As the wastewater is harmful to the environment and people, there are strict requirements for the emission of the wastewater. However, due to the difference in the raw materials, products, dyes, technology and equipment, the standards of the wastewater emission have too much items. It is developed by the national environmental protection department according to the local conditions and environmental protection requirements which is not fixed. It varies according to the situation in different regions. Therefore, the nature of emission targets is priorities of the points.

Serial number	Parameters	The Limits of Discharged Concentration	The Limits of Discharged Concentration for new Factory	The Special Limits of Discharged Concentration
1	COD	100mg/L	80 mg/L	60 mg/L
2	BOD	25 mg/L	20 mg/L	15 mg/L
3	pH	6~9	6~9	6~9
4	SS	70 mg/L	60 mg/L	20 mg/L
5	Chrominance	80	60	40
6	TN	20 mg/L	15 mg/L	12 mg/L
7	NH3-N	15 mg/L	12 mg/L	10 mg/L
8	TP	1.0 mg/L	0.5 mg/L	0.5 mg/L
9	S	1.0 mg/L	Can not be detected	Can not be detected
10	ClO2	0.5 mg/L	0.5 mg/L	0.5 mg/L
11	Cr6+	0.5 mg/L	Can not be detected	Can not be detected
12	Aniline	1.0 mg/L	Can not be detected	Can not be detected

Table 3. "Textile industry standards for water pollutants"

For printing and dyeing wastewater, the first consideration is the organic pollutants, color and heavy metal ions. Recently, as the lack of water, the recovery of wastewater should be considered. So the decolorization of the printing and dyeing wastewater increased heavily. The standards of printing and dyeing are different in different countries.Through access to the relevant information, the textile industry standards for water pollutants in China, Germany, U.S have been found.

1.4.1 Textile industry standards for water pollutants in China
For the emission standards of the textile dyeing wastewater in China, it is the very stringentstandards in the world. The emission standards for different indicators in textile industry standards for water pollutants in China have been shown in table 3 ("Discharge standard of water pollutants for dyeing and finishing of textile industry").

1.4.2 Textile industry standards for water pollutants in Germany
The emission standards for different indicators in textile industry standards for water pollutants in Germany have been shown in table 4 ("Discharge standard of water pollutants for dyeing and finishing of textile industry").

Serial number	Parameters	The Limits of Discharged Concentration
1	COD	160mg/L
2	BOD	25 mg/L
3	TP	2.0 mg/L
4	TN	20 mg/L
5	NH3-N	10 mg/L
6	Nitrite	1.0 mg/L

Table 4. Textile industry standards for water pollutants

The requirements of ammonia and total nitrogen are adjusted for the biochemical outflow at 12°C or above. Besides, the standard has also made the following emission requirements for the wastewater at the production stain.

There must not be in the wastewater:

1. Organic chlorine carriers (dyed acceleration)
2. Separation of chlorine bleach materials, except the sodium chlorite from the bleached synthetic fibers
3. The free chlorine after using sodium chlorite
4. Arsenic, mercury and their mixtures
5. Alkyl phenol as a bleaching agent （APEO）
6. Cr6 + compounds in the oxidizing of sulfur dyes and vat dyes
7. EDTA, DTPA, and phosphate in the water treatment softeners
8. Accumulation of chemicals, dyes and textile auxiliaries

1.4.3 Textile industry standards for water pollutants in U.S.

Printing and dyeing wastewater

It is the order for the printing and dyeing, including rinsing, dyeing, bleaching, washing, drying and other similar processes.

The requirements using BPT （best practical control tech.） which is published by EPA has shown in Table 5 ("Discharge standard of water pollutants for dyeing and finishing of textile industry").

Fabric printing and dyeing wastewater

It is adjusted for the fabric printing and dyeing wastewater, including bleaching, mercerization, dyeing, resin processing, washing, drying and so on.

The requirements using BPT (best practical control tech.) to treat the fabric printing and dyeing wastewater has been shown in Table 6 ("Discharge standard of water pollutants for dyeing and finishing of textile industry").

Serial number	Parameters	BPT	
		Maximum	Average of 30 days
		Kg/t(Fabric)	
1	BOD$_5$	22.4	11.2
2	COD	163.0	81.5
3	TSS	35.2	17.6
4	S	0.28	0.14
5	Phenol	0.14	0.07
6	Cr	0.14	0.07
7	pH	6.0~9.0	6.0~9.0

Table 5. Emission standards for gross printing and dyeing wastewater

Yarn printing and dyeing wastewater

It is adjusted for the Yarn printing and dyeing wastewater, including washing, mercerization, resin processing, dyeing and special finishing.

The requirements using BPT (best practical control tech.) to treat the yarn printing and dyeing wastewater has been shown in Table 7 ("Discharge standard of water pollutants for dyeing and finishing of textile industry").

Serial number	Parameters	BPT	
		Maximum	Average of 30 days
		Kg/t(Fabric)	
1	BOD$_5$	5.0	2.5
2	COD	60	30
3	TSS	21.8	10.9
4	S	0.20	0.10
5	Phenol	0.10	0.05
6	Cr	0.10	0.05
7	pH	6.0~9.0	6.0~9.0

Table 6. Emission standards for fabric printing and dyeing wastewater

Serial number	Parameters	BPT	
		Maximum	Average of 30 days
		Kg/t(Fabric)	
1	BOD$_5$	6.8	3.4
2	COD	84.6	42.3
3	TSS	17.4	8.7
4	S	0.24	0.12
5	Phenol	0.12	0.06
6	Cr	0.12	0.06
7	pH	6.0~9.0	6.0~9.0

Table 7. Emission standards for yarn printing and dyeing wastewater

2. Textile dyeing wastewater treatment processes

The textile dyeing wastewater has a large amount of complex components with high concentrations of organic, high-color and changing greatly characteristics. Owing to their high BOD/COD, their coloration and their salt load, the wastewater resulting from dyeing cotton with reactive dyes are seriously polluted. As aquatic organisms need light in order to develop, any deficit in this respect caused by colored water leads to an imbalance of the ecosystem. Moreover, the water of rivers that are used for drinking water must not be colored, as otherwise the treatment costs will be increased. Obviously, when legal limits exist (not in all the countries) these should be taken as justification. Studies concerning the feasibility of treating dyeing wastewater are very important (C. All`egre et al.,2006).

In the past several decades, many techniques have been developed to find an economic and efficient way to treat the textile dyeing wastewater, including physicochemical, biochemical, combined treatment processes and other technologies. These technologies are usually highly efficient for the textile dyeing wastewater.

2.1 Physicochemical wastewater treatment

Wastewater treatment is a mixture of unit processes, some physical, others chemical or biological in their action. A conventional treatment process is comprised of a series of individual unit processes, with the output (or effluent) of one process becoming the input (influent) of the next process. The first stage will usually be made up of physical processes. Physicochemical wastewater treatment has been widely used in the sewage treatment plant which has a high removal of chroma and suspended substances, while it has a low removal of COD. The common physicochemical methods are shown as followed.

2.1.1 Equalization and homogenization

Because of water quality highly polluted and quantity fluctuations, complex components, textile dyeing wastewater is generally required pretreatment to ensure the treatment effect and stable operation.

In general, the regulating tank is set to treat the wastewater. Meantime, to prevent the lint, cotton seed shell, and the slurry Settle to the bottom of the tank, it's usually mixed the wastewater with air or mechanical mixing equipment in the tank. The hydraulic retention time is generally about 8 h.

2.1.2 Floatation

The floatation produces a large number of micro-bubbles in order to form the three-phase substances of water, gas, and solid. Dissolved air under pressure may be added to cause the formation of tiny bubbles which will attach to particles. Under the effect of interfacial tension, buoyancy of bubble rising, hydrostatic pressure and variety of other forces, the microbubble adheres to the tiny fibers. Due to its low density, the mixtures float to the surface so that the oil particles are separated from the water. So, this method can effectively remove the fibers in wastewater.

2.1.3 Coagulation flocculation sedimentation

Coagulation flocculation sedimentation is one of the most used methods, especially in the conventional treatment process.

Active on suspended matter, colloidal type of very small size, their electrical charge give repulsion and prevent their aggregation. Adding in water electrolytic products such as aluminum sulphate, ferric sulphate, ferric chloride, giving hydrolysable metallic ions or organic hydrolysable polymers (polyelectrolyte) can eliminate the surface electrical charges of the colloids. This effect is named coagulation. Normally the colloids bring negative charges, so the coagulants are usually inorganic or organic cationic coagulants (with positive charge in water).

The metallic hydroxides and the organic polymers, besides giving the coagulation, can help the particle aggregation into flocks, thereby increasing the sedimentation. The combined action of coagulation, flocculation and settling is named clariflocculation.

Settling needs stillness and flow velocity, so these three processes need different reactions tanks. This processes use mechanical separation among heterogeneous matters, while the dissolved matter is not well removed (clariflocculation can eliminate a part of it by absorption into the flocks). The dissolved matter can be better removed by biological or by other physical chemical processes (Sheng.H et al.,1997) .

But additional chemical load on the effluent (which normally increases salt concentration) increases the sludge production and leads to the uncompleted dye removal.

2.1.4 Chemical oxidation

Chemical operations, as the name suggests, are those in which strictly chemical reactions occur, such as precipitation. Chemical treatment relies upon the chemical interactions of the contaminants we wish to remove from water, and the application of chemicals that either aid in the separation of contaminants from water, or assist in the destruction or neutralization of harmful effects associated with contaminants. Chemical treatment methods are applied both as stand-alone technologies and as an integral part of the treatment process with physical methods (K.Ranganathan et al.,2007).Chemical operations can oxidize the pigment in the printing and dyeing wastewater as well as bleaching the effluent. Currently, Fenton oxidation and ozone oxidation are often used in the wastewater treatment.

2.1.4.1 Fenton reaction

Oxidative processes represent a widely used chemical method for the treatment of textile effluent, where decolourisation is the main concern. Among the oxidizing agents, the main chemical is hydrogen peroxide (H_2O_2), variously activated to form hydroxyl radicals, which are among the strongest existing oxidizing agents and are able to decolourise a wide range of dyes.

A first method to activate hydroxyl radical formation from H_2O_2 is the so called Fenton reaction, where hydrogen peroxide is added to an acidic solution (pH=2-3) containing Fe^{2+} ions.

Fenton reaction is mainly used as a pre-treatment for wastewater resistant to biological treatment or/and toxic to biomass. The reaction is exothermic and should take place at temperature higher than ambient. In large scale plants, however, the reaction is commonly carried out at ambient temperature using a large excess of iron as well as hydrogen peroxide. In such conditions ions do not act as catalyst and the great amount of total COD removed has to be mainly ascribed to the $Fe(OH)_3$ co-precipitation. The main drawbacks of the method are the significant addition of acid and alkali to reach the required pH, the necessity to abate the residual iron concentration, too high for discharge in final effluent, and the related high sludge production (Sheng.H et al.,1997).

2.1.4.2 Ozone oxidation

It is a very effective and fast decolourising treatment, which can easily break the double bonds present in most of the dyes. Ozonation can also inhibit or destroy the foaming properties of residual surfactants and it can oxidize a significant portion of COD.

Moreover, it can improve the biodegradability of those effluents which contain a high fraction of nonbiodegradable and toxic components through the conversion (by a limited oxidation) of recalcitrant pollutants into more easily biodegradable intermediates. As a further advantage, the treatment does increase neither the volume of wastewater nor the sludge mass.

Full scale applications are growing in number, mainly as final polishing treatment, generally requiring up-stream treatments such as at least filtration to reduce the suspended solids contents and improve the efficiency of decolourisation. Sodium hypochlorite has been widely used in the past as oxidizing agent. In textile effluent it initiates and accelerates azo bond cleavage. The negative effect is the release of carcinogenic aromatic amines and otherwise toxic molecules and, therefore, it should not be used (Sheng.H et al.,1997).

2.1.5 Adsorption

Adsorption is the most used method in physicochemical wastewater treatment, which can mix the wastewater and the porous material powder or granules, such as activated carbon and clay, or let the wastewater through its filter bed composed of granular materials.

Through this method, pollutants in the wastewater are adsorbed and removed on the surface of the porous material or filter.

Commonly used adsorbents are activated carbon, silicon polymers and kaolin. Different adsorbents have selective adsorption of dyes. But, so far, activated carbon is still the best adsorbent of dye wastewater. The chroma can be removed 92.17% and COD can be reduced 91.15% in series adsorption reactors, which meet the wastewater standard in the textile industry and can be reused as the washing water. Because activated carbon has selection to adsorb dyes, it can effectively remove the water-soluble dyes in wastewater, such as reactive dyes, basic dyes and azo dyes, but it can't adsorb the suspended solids and insoluble dyes. Moreover, the activated carbon can not be directly used in the original textile dyeing wastewater treatment, while generally used in lower concentration of dye wastewater treatment or advanced treatment because of the high cost of regeneration.

2.1.6 Membrane separation process

Membrane separation process is the method that uses the membrane's micropores to filter and makes use of membrane's selective permeability to separate certain substances in wastewater. Currently, the membrane separation process is often used for treatment of dyeing wastewater mainly based on membrane pressure, such as reverse osmosis, ultrafiltration, nanofiltration and microfiltration. Membrane separation process is a new separation technology, with high separation efficiency, low energy consumption, easy operation, no pollution and so on. However, this technology is still not large-scale promoted because it has the limitation of requiring special equipment, and having high investment and the membrane fouling and so on (K.Ranganathan et al.,2007) .

2.1.6.1 Reverse osmosis

Reverse osmosis membranes have a retention rate of 90% or more for most types of ionic compounds and produce a high quality of permeate. Decolorization and elimination of chemical auxiliaries in dye house wastewater can be carried out in a single step by reverse osmosis. Reverse osmosis permits the removal of all mineral salts, hydrolyzed reactive dyes, and chemical auxiliaries. It must be noted that higher the concentration of dissolved salt, the more important the osmotic pressure becomes; therefore, the greater the energy required for the separation process (B. Ramesh Babu et al.,2007) .

2.1.6.2 Nanofiltration

Nanofiltration has been applied for the treatment of colored effluents from the textile industry. Its aperture is only about several nanometers, the retention molecular weaght by which is about 80-1000da, A combination of adsorption and nanofiltration can be adopted for the treatment of textile dye effluents. The adsorption step precedes nanofiltration, because this sequence decreases concentration polarization during the filtration process, which increases the process output. Nanofiltration membranes retain low molecular weight organic compounds, divalent ions, large monovalent ions, hydrolyzed reactive dyes, and dyeing auxiliaries. Harmful effects of high concentrations of dye and salts in dye house effluents have frequently been reported. In most published studies concerning dye house effluents, the concentration of mineral salts does not exceed 20 g/L, and the concentration of dyestuff does not exceed 1.5 g/L. Generally, the effluents are reconstituted with only one dye, and the volume studied is also low. The treatment of dyeing wastewater by nanofiltration represents one of the rare applications possible for the treatment of solutions with highly c oncentrated and complex solutions (B. Ramesh Babu et al.,2007).

A major problem is the accumulation of dissolved solids, which makes discharging the treated effluents into water streams impossible. Various research groups have tried to develop economically feasible technologies for effective treatment of dye effluents. Nanofiltration treatment as an alternative has been found to be fairly satisfactory. The technique is also favorable in terms of environmental regulating.

2.1.6.3 Ultrafiltration

Ultrafiltration whose aperture is only about 1nm-0.05 μ m, enables elimination of macromolecules and particles, but the elimination of polluting substances, such as dyes, is never complete. Even in the best of cases, the quality of the treated wastewater does not permit its reuse for sensitive processes, such as dyeing of textile. So the retention molecular weaght is range from 1000-300000da, . Rott and Minke (1999) emphasize that 40% of the water treated by ultrafiltration can be recycled to feed processes termed "minor" in the textile industry (rinsing, washing) in which salinity is not a problem. Ultrafiltration can only be used as a pretreatment for reverse osmosis or in combination with a biological reactor (B. Ramesh Babu et al.,2007) .

2.1.6.4 Microfiltration

Microfiltration whose aperture is about 0.1-1 μ m is suitable for treating dye baths containing pigment dyes, as well as for subsequent rinsing baths. The chemicals used in dye bath, which are not filtered by microfiltration, will remain in the bath. Microfiltration can also be used as a pretreatment for nanofiltration or reverse osmosis (B. Ramesh Babu et al.,2007).

Textile wastewater contains large amounts of difficult biodegradable organic matter and inorganic. At present , many factories have adopted physicochemical treatment process. Some typical physicochemical treatment process is shown in Table 8 and Fig. 2 (Kangmei Zeng et al.,2005).

2.2 Biological wastewater treatment method

The biological process removes dissolved matter in a way similar to the self depuration but in a further and more efficient way than clariflocculation. The removal efficiency depends upon the ratio between organic load and the bio mass present in the oxidation tank, its temperature, and oxygen concentration.

The bio mass concentration can increase, by aeration the suspension effect but it is important not to reach a mixing energy that can destroy the flocks, because it can inhibit the following settling.

Normally, the biomass concentration ranges between 2500-4500 mg/l, oxygen about 2 mg/l. With aeration time till 24 hours the oxygen demand can be reduced till 99%.

According to the different oxygen demand, biological treatment methods can be divided into aerobic and anaerobic treatment. Because of high efficiency and wide application of the aerobic biological treatment, it naturally becomes the mainstream of biological treatment.

2.2.1 Aerobic biological treatment

According to the oxygen requirements of the different bacteria, the bacteria can be divided into aerobic bacteria, anaerobic bacteria and facultative bacteria. Aerobic biological treatment can purify the water with the help of aerobic bacteria and facultative bacteria in the aerobic environment. Aerobic biological treatment can be divided into two major categories: activated sludge process and biofilm process.

Name	Dyes and Additives in sewage	Water Quantity (t/d)	Main Process	Amount of Coagulant (mg/L)	Water Qualify		Treatment Efficiency	
					Color (times)	COD (mg/L)	Color (%)	COD (%)
A Knitting Mill in Kunming Yunnan	Naphthol, Direct dye, Acidic dye, Reactive dye	1000	wastewater → pump (↓ PAC) → mixed reaction tank / effluent ← sedimentation tank	60-80	70-120	267	>90	60
A Printing and Dyeing Mill in Shanghai	Vat dye, Naphthol, Dope	Pilot scale test	wastewater → regulating tank → pump → dissoved vessel → floatation tank → effluent	400	400	600-800	80-90	60
A Printing and Dyeing Mill in Beijing	Acidic dye, Disperse dye, Reactive dye Sulfide dye	120	wastewater → regulating tank → coagulatio tube sedimentation tank → floatation tank → sand filter tank → effluent	—	174-347	228-352	97.41	74.69
A Silk and Dyeing Mill in Shaoxing, zhejiang	Disperse dye, Reactive dye, Direct dye, Sulfide dye, Acidic dye	500	wastewater → regulating tank → reaction tank → coagulation tank → tube sedimentation tank → effluent tank	$FeSO_4$: 0.7kg/t sewage, Lime: 0.38kg/t sewage.	720-830	1114-1153	92	55-59

Table 8. Physicochemical treatment instance of textile dyeing wastewater

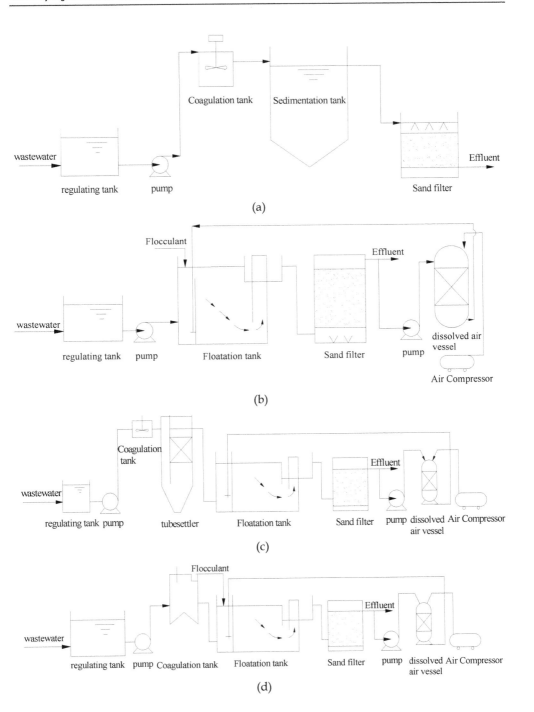

(a)

(b)

(c)

(d)

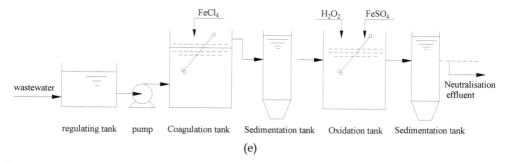

(e)

Fig. 2. The physicochemical treatment process in textile dyeing wastewater

2.2.1.1 Activated sludge process

Activated sludge is a kind of floc which is mainly comprised of many microorganisms, which has strong decomposition and adsorption of the organics, so it is called "activated sludge". The wastewater can be clarified and purified after the separation of activated sludge. Activated sludge process is based on the activated sludge whose main structure is the aeration tank.

Activated sludge Process is an effective method. As long as according to the scientific laws, after getting some experiences, higher removal efficiency can be got. In present treatment, the oxidation ditch and SBR process are the commonly used activated sludge process.

Oxidation ditch process

Oxidation ditch is a kind of biological wastewater treatment technology, which developed by the netherlands health engineering research institute in the 50's of last century. It is a variant of activated sludge, which is a special form of extended aeration. The oxidation ditch plan was shown in Fig. 3. And the biological sewage treatment process which mainly composed by the oxidation ditch, as was shown in Fig. 4.

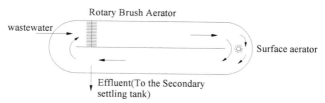

Fig. 3. Oxidation ditch plan

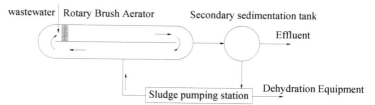

Fig. 4. The oxidation ditch for the biological treatment process

The oxidation ditch is generally consisted of the ditch body, aeration equipment, equipment of the water in or out, diversion and mixing equipment. The shape of the ditch body is usually ring; it also can be rectangular, L-shaped, round or other shape. The side shape of the oxidation ditch generally is rectangular or trapezoidal. Wastewater, activated sludge, and various microorganisms are mixed in a continuous loop ditch in order to complete nitrification and denitrification. Oxidation ditch has the characteristics of completely mixed, plug-flow and oxidation tank. Since oxidation ditch has long hydraulic retention time(HRT), low organic loading and long sludge age, compared to conventional activated sludge process, the equalization tank, primary sedimentation tank, sludge digestion tank can be omitted. The secondary sedimentation tank also can be omitted in some processes. It has many characteristics, such as high degree of purification, impact resistance, stable, reliable, simple, easy operation and management, easy maintenance, low investment and energy consumption. Oxidation ditch forms aerobic zone, anoxic zone and the anaerobic zone in space, which has a good function of denitrification.

Sequencing Batch Reactor Activated Sludge process

Sequencing Batch Reactor Activated Sludge Process (SBR Process) is a reform process from activated sludge, which is a new operating mode. Its operation is mainly composed of five processes: ①inflow; ②reaction; ③sedimentation; ④outflow; ⑤standby. The reaction process plan has been showed in Fig. 5.

SBR treatment process not only has a high removal rate of COD, but also has a high removal efficiency of color. Compared to the traditional methods, using SBR process has the following advantages (Li-yan Fu et al.,2001):

i.   It has great resistance to shock loading. The reserved water can effectively resist the impact of water and organic matter.

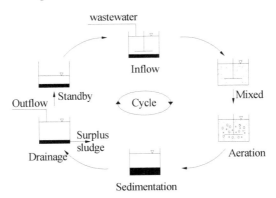

Fig. 5. Reaction process of SBR

ii. Flexible control. The run time, the total residence time and gas supply of each stage can be adjusted according to the quantity of inflow or outflow.

iii. The activated sludge has good traits and low sludge production rate. Since the original water contains many organic matters, which is suitable for the bacteria to grow, the sludge is in good characters. In the standby period, the sludge is in the endogenous respiration phase, so the sludge's yield is low.

iv. It has less processing equipment, simple structure ⸲ and it is easy to be operated.

items	wastewater	Effluent	Removal rate
CODcr(mg/L)	689	61.2	91.2
Color(times)	800	50	93.8
SS(mg/L)	384.3	32.6	91.5

Table 9. The removal efficiency of various indicators in this dyeing factory

There is a dyeing factory in Jiangsu, which produces 60,000 tons wastewater each year. The wastewater mainly comes from dyeing and rinse. The major pollutant in wastewater is seriously exceeded, such as COD, color and SS. The wastewater treatment process which is used in this factory has been shown in Fig.2.5. Through this process, its various indicators have been shown in table 9(Haidong Jia,2003).

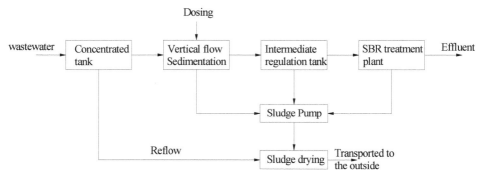

Fig. 6. The wastewater treatment process based on SBR

2.2.1.2 Biofilm

The biofilm process is a kind of biological treatment that making the numerous microorganisms to attach to some fixed object surface, while letting the wastewater flow on its surface to purify it by contact. The main types of the biofilm process are biological contact oxidation, rotating biological contactors and biological fluidized bed.

Biological contact oxidation

Biological contact oxidation is widely used in the dyeing wastewater treatment. The main feature of the process is to set fillers in the aeration tanks, so that it has the characteristics of activated sludge and biofilm. The wastewater in oxidation tank contains a certain amount of activated sludge, while the fillers are covered with a large number of biofilm. When the wastewater contact with the fillers, it can be purified under the function of aerobic microorganisms.

1. The main features of the biological contact oxidation tank are as following:
    i.   It has a efficient purification, a short processing time as well as a good and stable water quality.
    ii. It has a higher ability to adapt the impact load. Under the intermittent operating condition, it is still able to maintain good treatment effect. To uneven drainage enterprises, it has more actual significance.
    iii. Its operation is simple, convenient and easy maintenance. It don't have sludge return, sludge bulking phenomenon or any filter flies.

iv. It produce less sludge and the sludge particles is larger ,which is easier to be sedimentated

2.   The form of biological contact oxidation

According to the different direction of the influent and aeration, the design form of the biological contact oxidation can be divided into two types: one is the same flow contact process which makes the wastewater and air flow into the contact tank bottom (Fig. 7). This process can guarantee the volume load about 4.5 kgBOD/m³. Another is the reverse flow contact process which let the air flow into the bottom, and let the wastewater into the top(Fig. 8). This process can guarantee the volume load about 8 kgBOD/m³.

3.   The choice of filler

The filler is not only related to the treatment effect, but also affects the project investment. The surface area, bio-adhesion and whether easily blocked of the filler are undoubtedly the most important conditions, but the economy is also an important factors. The filler in the oxidation tank is the proportion of investment accounted for relatively large, so the price is often the first consideration. For example, the technical performance of some filler is slightly worse, but the price is cheap. The contact time can be increased appropriately. Commonly used fillers are honeycomb packing, corrugated board packing equipment, soft and semi-soft filler.

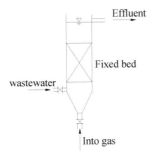

Fig. 7. The schematic diagram of the same flow contact

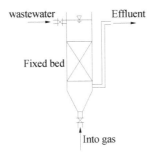

Fig. 8. The schematic diagram of the reverse flow contact

Rotating Biological Contactor

Rotating Biological Contactor is an efficient sewage treatment plant developed on the basis of the original biological filter. It is constituted by a series of closed disks which are fixed on a horizontal axis (Fig. 9). The disks are made of lightweight materials, such as hard plastic plate, glass plate etc.The disk diameter is about 1-3m generally. Nearly half of the disk area is submerged in the sewage of the oxidation tank, but the upper half is exposed to the air. The rotating horizontal axis which is driven by the Rotating device makes the disk rotating slowly. Due to the disk's rotating, the sewage in the oxidation tank is completely mixed. There is a layer of biofilm on the disk surface, when the rotating disk immersed in the sewage inside of the oxidation tank, the organic matter in the sewage would be adsorbed by the biofilm on the disk. When the rotating disk rotate to the air, the water film which is brought up by the disk will drip down along the biofilm surface, at the same time, the oxygen in the air will dissolve into the water film constantly. Under the catalysis of enzyme, though absorbing the dissolved oxygen in the water film, microorganism can oxidate and decompose the organic matter in the sewage and excrete the metabolite at the same time. In the process of disk rotation, the biofilm on the disk get in touch with the sewage and the air constantly alternating, completing the process of adsorption-oxidation-oxidative decomposition continuously to purify the sewage. The advantages of the Biological Rotating Contactor are power saving, large shock load capability, no sludge return, little sludge generated, little noise, easy maintenance and management and so on. On the contrary, the shortcomings are the large area and needing the maintain buildings.

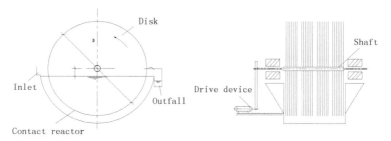

Fig. 9. The schematic diagram of biological rotating disk

The main parameters of affecting the process performance are rotate speed, sewage residence time, reactive tank stage, disk submergence and temperature.

Towards the low consistency sewage whose BOD consistency is under 300 mg/L, when the rotate speed is under 18 m/min, the disposal efficiency would enhance with the rotate speed increase, otherwise, it would not be any improving. Toward high-BOD consistency sewage, increasing the rotate speed is equivalent to increase the contact, oxidation and stirring. So, it improves the efficiency. On the other hand, increasing the rotate speed increase the energy consumption shapely, so we must do full economic comparison between increasing energy consumption and increasing land acreage.

Biological fluidized bed

Fluidized bed process is also called suspended carrier biofilm process, which is a new efficient sewage treatment process. The method which adopts the solid particles fluidization technology can keep the whole system at a fluidized state to enhance the contact of solid particles and fluid and achieve the purpose of sewage purification. The substance of this method is that using activated carbon, sand, anthracite or other particles whose diameter is less than 1mm as the carrier and filling in the container. Through pulse water, the carrier is made into fluidization and covered by the biofilm on its surface. Due to the small size of the carrier, it has a great surface area in unit volume (the carrier surface area can reach 2000-3000 m^2/m^3), and can maintain a high level of microbial biomass. The treatment efficiency is about 10-20 times higher than conventional activated sludge process. It can remove much organic matter in short time, but the land acreage is only about 5% of the common activated sludge process. Therefore, in the treatment of high consistency organic wastewater in the textile dyeing wastewater, we can consider adopting this suspended carrier fluidized bed process.

2.2.2 Anaerobic biological treatment
Anaerobic biological treatment process is a method that make use of the anaerobic bacteria decompose organic matter in anaerobic conditions. This method was first used for sludge digestion. In recent years it was gradually used in high concentration and low concentration organic wastewater treatment. In textile industry, there are many types of high concentration organic wastewater, such as wool washing sewage, textile printing and dyeing wastewater etc., which the organic matter content of it is as high as 1000 mg/L or more, the anaerobic wastewater treatment process can achieve good results. The anaerobic-aerobic treatment process is usually adopted in actual project that is using anaerobic treatment to treat high concentration wastewater, and using aerobic treatment to treat low concentration wastewater. Currently, the hydrolysis acidification process is the main anaerobic treatment process, which can increases the biodegradability of the sewage to facilitate the following biological treatment process.

The hydrolysis acidification process is the first two stages of the anaerobic treatment. Through making use of the anaerobic bacteria and facultative bacteria, the macromolecule, heterocyclic organic matter and other difficult biodegradable organic matter would be decomposed into small molecular organic matter, thereby enhancing the biodegradability of the wastewater and destructing the colored groups of dye molecules to remove part of the color in wastewater. More importantly, due to the molecular structure of the organic matter and colored material or the chromophore has been changed by the anaerobic bacteria, it's easy to decompose and decolor under the aerobic conditions, which improve the decolorization effect of the sewage. Operating data shows that the pH value of the effluent from hydrolysis tank usually decrease 1.5 units. The organic acid which is produced in hydrolysis can effectively neutralize some of the alkalinity in wastewater, which can make the pH value of sewage drop to about 8 to provide a good neutral environment for following aerobic treatment. Currently, the anaerobic digestion process is a essential measure in the biological treatment of textile dyeing wastewater.

In addition, there are many other processes used in textile dyeing wastewater treatment currently, such as upflow anaerobic sludge bed(UASB), upflow anaerobic fluidized bed (UABF), anaerobic baffled reactor (ABR) and anaerobic biological filter and so on.

2.3 Biochemical and physicochemical combination processes
In recent years, as the application of new technologies in textile and dyeing industry, a large number of difficult biodegradable organic matter such as PVA slurry, surface active agents and new additives enter into the dyeing wastewater, which result in the high concentration of the organic matter, complex and changeable composition and the obvious reduction of the biodegradability. The COD_{Cr} removal rate of the simple aerobic activated sludge process which was used to treat the textile dyeing wastewater has decreased from 70% to 50%, and the effluent can not meet the discharge standards. More seriously, quite a number of sewage treatment facilities can't normally operate even stop running. Therefore, the biochemical

and physicochemical combination processes has been gradually developed. And its application is increasingly widespread (Sheng-Jie You et al.,2008). The types of the combination process are various, and the main adoptions currently are as following:

2.3.1 Hydrolytic acidification-contact oxidation-air floatation process

This combination process is a typical treatment process of the textile dyeing wastewater, which is widely used (The process flow diagram is shown in Fig. 10).The wastewater firstly flows through the bar screen, in order to remove a part of the larger fibers and particles, and Then flows into the regulating tank. After well-distributed through a certain amount of time, the sewage flows into the hydrolysis acidification tank to carry out the anaerobic hydrolysis reaction. The reaction mechanism is making use of the anaerobic hydrolysis and acidification reaction of the anaerobic fermentation to degrade the insoluble organic matter into the soluble organic matter by controlling the hydraulic retention time. At the same time, through cooperating with the acid bacteria, the macromolecules and difficult biodegradable organic matter would be turned into biodegradable small molecules, which provide a good condition for the subsequent biological treatment. Next, the sewage enters into the biological contact oxidation tank. After the biochemical treatment, the wastewater directly enters into the flotation tank for flotation treatment, which is adopted the pressurized full-dissolved air flotation process. The polymer flocculants added in flotation tank react with the hazardous substances, which can condense the hazardous substances into tiny particles. Meanwhile, sufficient air is dissolved in the wastewater. And then the pressure suddenly is released to produce uniformly fine bubbles, which would adhere to the small particles. The density of the formation is less than $1kg/m^3$, which can make the formation float and achieve the separation of the solid and liquid.

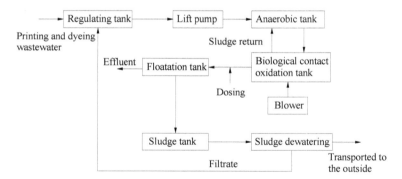

Fig. 10. The process flow diagram

The anaerobic hydrolysis acidification tank equipped with semi-soft padding and the biological contact oxidation tank equipped with the new SNP-based filler. The following physicochemical treatment uses the dosing flotation tank, which has four characteristics. Firstly, the deciduous biofilm and suspended solids removal rate can reach 80% to 90%. Secondly, the color removal rate can reach 95%. Thirdly, the hydraulic retention time in the flotation tank is short, which is only about 30 min, while the precipitation tank is about 1.15 h to 2 h, so the volume and area of the flotation tank is small. Finally, the sludge moisture content is low, only about 97% to 98%, which can be directly dewatered. But the flotation treatment need an additional air compressor, pressure dissolved gas cylinders, pumps and other auxiliary system. The operation and management is also relatively complicated.

After the treatment of this process, the COD_{Cr} removal rate can be up to 95% or more. The actual effluent quality is about: pH=6~9, color<100times, SS<100mg/L, BOD5<50mg/L, COD_{Cr}<150mg/L (Honglian Li, 2006).

2.3.2 Anaerobic-aerobic-biological carbon contacts

The treatment process is a mature and widely used process in wastewater treatment in recent years (the process shown in Fig. 11). The anaerobic treatment here is not the traditional anaerobic nitrification, but the hydrolysis and acidification. The purpose is aiming at degrading some poorly biodegradable polymer materials and insoluble material in textile dyeing wastewater to small molecules and soluble substances by hydrolysis and acidification, meanwhile, improving the biodegradability and BOD_5/COD_{Cr} value of the wastewater in order to create a good condition for the subsequent aerobic biological treatment. At the same time, all sludge generated in the aerobic biological treatment return into the anaerobic biological stage through the sedimentation tank. Because of the sludge in the anaerobic biochemical stage has sufficient hydraulic retention time (8h~10h) to carry out anaerobic digest thoroughly, The whole system would not discharge sludge, that is the sludge achieve its own balance (Note: only a small amount of inorganic sludge accumulate in the anaerobic stage, but do not have to set up a special sludge treatment plant).

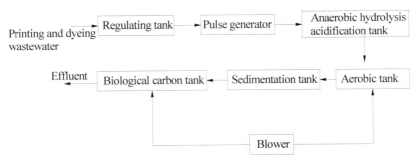

Fig. 11. The process flow diagram

Anaerobic tank and aerobic tank are both installed media, which is a biofilm process. Biological carbon tank is filled with activated carbon and provided oxygen, which has the characteristics both of suspended growth method and fixation growth method. The function of pulse water is mixing in the anaerobic tank. The hydraulic retention time of various parts is about:

Regulating tank: 8h~12h; anaerobic biochemical tank: 8h~10h
Aerobic biochemical tank: 6 ~8h; biological carbon tank: 1h~2h
Pulse generator interval: 5min~10min.

According to the textile dyeing wastewater standard ($COD_{Cr}\leq1000mg/L$), the effluent can achieve the national emission standards, which can be reused through further advanced treatment. For the five years operation project, the results show that the operation is normal, the treatment effect is steady, there is no efflux of sludge and the sludge was not found excessive growth in the anaerobic biochemical tank.

2.3.3 Coagulation-ABR-oxidation ditch process
The treatment has been adopted widely currently, such as a textile dyeing wastewater treatment plant in Jiangmen of Guangdong Province (the process is shown in Fig. 12).
The characteristics of the textile dyeing plant effluent are the variation in water, the higher of the alkaline, color and organic matter concentration , and the difficulty of the degradation (BOD_5/COD_{Cr} value is about 0.25). The workshop wastewater enter into the regulating tank by pipe network to balance the quantity and quality, after wiping off the large debris by the bar screen before the regulating tank. The adjusted wastewater flow into the coagulation reaction tank, at the same time, the $FeSO_4$ solution was added into it to carry out chemical reaction. Finally, the effluent flows into the primary sedimentation tank for spate separation, meanwhile enhancing the BOD_5/COD_{Cr} ratio.

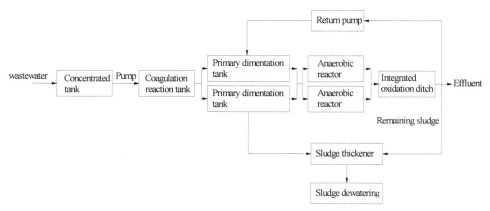

Fig. 12. The process flow diagram

The effluent of primary sedimentation tank flows into the anaerobic baffled reactor (ABR) by gravity. After the anaerobic hydrolysis reaction, it enters into the integrated oxidation ditch for aerobic treatment, and then goes into the secondary sedimentation tank for spate separation. The upper liquid was discharged after meeting the standards, but the settled sludge was returned to the return sludge tank, most of which was returned to ABR anaerobic tank by pump. The remaining sludge was pumped to the sludge thickening tank for concentration.
Since the using of this sewage treatment process, the treatment effect is stable. The removal rate of each index is high, and the operation situation of each unit is shown in Table10 (Wu Zhimin&Zhangli, 2009).

Process	PH	CODcr (mg/L)	BOD$_5$ (mg/L)	SS (mg/L)	Color (times)
Wastewater	9-13	1800-2000	400-500	250-350	500
Coagulation effluent	6-9	1327	344	157	102
ABR effluent	6-9	532	292	94	48
Aerobic secondly sedimentation effluent	6-9	80	15	14	30

Table 10. Average Quality of Each Process Effluent

2.3.4 UASB-aerobic-physicochemical treatment process

For the high pH value dyeing wastewater treatment systems, we should adjust the pH value to the range from 6 to 9before entering the treatment system. While towards the difficultly biodegradable textile dyeing wastewater, effective measures should be taken to increase the biodegradability. At the same time, the volume of the regulating tank must be increased (this is repeatedly demonstrated in engineering practice) to guarantee the water quantity, quality and color achieving a relatively uniform towards changeable dyeing wastewater. Good inflow condition can be created for following process though this method. Therefore, we put forward the "UASB-aerobic-physicochemical method" treatment process in the actual project, and the technological process is shown in Fig. 13.

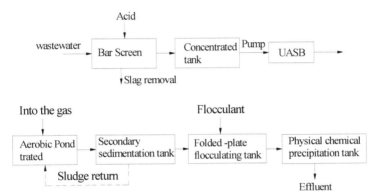

Fig. 13. The process flow diagram

At present, the treatment process has been applied in a number of textile dyeing factory, such as a Jiangyin textile dyeing factory, which sewage component is complex. Before the sewage enters into the regulating tank, the suspended particles in which must be removed by bar screen, at the same time, an appropriate amount of acid was added to adjust the pH value of the sewage. Then adjusting the water quantity and quality to make it uniformed. After pretreatment, the textile dyeing wastewater carries out anaerobic reaction firstly in UASB reactor to improve the biodegradability of wastewater and the decolorization rate, and then flows into the aerobic tank. In the aerobic tank, the organic matter in sewage is removed. Finally, in order to ensure the effective removal of the suspended particles particular the activated sludge, some flocculants was added in the sedimentation tank to improve its effect. Through this sewage treatment process, the removal rate of pollutants is shown in table 11 (Kerong Zhang, 2007).

Items	(raw water) regulating tank	Biochemical treatment system		Physicochemical treatment system	
		Effluent	Removal rate	Effluent	Removal rate
PH	8-12	7-8		6-9	
CODcr(mg/L)	1000-2000	100-200	90	≤100	50
BOD$_5$(mg/L)	300-600	15-30	95	≤30	
Color(times)	100-600	60	80	≤40	35

Table 11. The removal of the processing units

2.4 Cutting-edge treatment process
2.4.1 Photochemical oxidation
Photochemical oxidation has many advantages of the mild reaction conditions (ambient temperature and pressure), powerful oxidation ability and fast speed, etc. It can be divided into 4 kinds, which are light decomposition, photoactivate oxidation, optical excitation oxidation and photocatalysis oxidation. Among them, the photocatalysis oxidation has been more researched and applied currently.

This technology can effectively destroy a lot of organic pollutants whose structure is stable and difficult to biologically degrade. Compared with the physical treatment in traditional wastewater treatment process, the most obvious advantages of this technology are significant energy efficiency, completely pollutants degradation and so on. Almost all of the organic matter can be completely oxidized to CO_2, H_2O and other simple inorganic substances under the light catalyst. However, towards high concentration wastewater, the effect of the photocatalysis oxidation process is not ideal. The research about photocatalysis oxidation degradation of dye mainly focused on the study of photocatalyst.

2.4.2 Electrochemical oxidation
The mechanism of the electrochemical process treating dyeing wastewater is making use of electrolytic oxidation, electrolytic reduction, electrocoagulation or electrolytic floating destruct the structure or the existence state to make it bleached. It has the advantages of small devices, small area covering, operation and management easily, higher COD_{Cr} removal rate and good bleaching effect, but the precipitation and the consumption of electrode material is great, and the operating cost is high. The traditional electrochemical methods can be divided into power flocculation, electrical float, electro-oxidation, micro-electrolysis and the electrolysis method. With the development of electrochemical technologies and the appearing of a variety of high-efficiency reactor, the cost of treatment will decrease largely. Electro-catalytic advanced oxidation process (AEOP) is a new advanced oxidation technology developed recently. Because of its high efficiency, easy operation, and environmental friendliness, it has attracted the attention of researchers. Under normal temperature and pressure, it can produce hydroxyl radicals directly or indirectly through the reactions in the catalytic activity electrode, thus the degradation of the difficultly biodegradable pollutants is effective. It is one of the main directions in future research.

2.4.3 Ultrasonic technology
Using ultrasonic technology can degrade chemical pollutants, especially the refractory organic pollutants in water. It combines the characteristics of advanced oxidation technology, incineration, supercritical water oxidation and other wastewater treatment technologies. Besides the degradation conditions are mild, degradation speed is fast and application widely, it can also use individually or combined with other water treatment technologies. The principle of this method is that the sewage enters into the air vibration chamber after being added the selected flocculants in regulating tank. Under the intense oscillations in nominal oscillation frequency, a part of organic matter in wastewater is changed into small organic molecule by destructing its chemical bonds. The flocculants flocculation rapidly companied with the color, COD_{Cr} and the aniline concentration was fall under the accelerating thermal motion of water molecules, which play the role of reducing organic matter concentration in wastewater. At present, the ultrasonic technology in the research of water treatment has achieved great achievements, but most of them are still confined to laboratory research level.

2.4.4 High energy physical process

High energy physical process is a new wastewater treatment technology. When the high-energy particle beam bombard aqueous solution, the water molecules would come up with excitation and ionization, produce ions, excited molecules, secondary electrons. Those products would interact with each other before spreading to the surrounding medium. It would produce highly reactive HO• radicals and H atoms, which would react with organic matter to degrade it. The advantages of using high-energy physics process treat dyeing wastewater are the small size of the equipment, high removal rate and simple operation. However, the device used to generate high energy particle is expensive, technically demanding is high, energy consumption is big, and the energy efficiency is low and so on. Therefore, it needs a lot of research work before put into actual project.

3. Conclusions and recommendations

Current, the textile dyeing wastewater is one of the most important source of pollution. The type of this wastewater has the characteristics of higher value of color, BOD and COD, Complex composition, large emission, widely distributed and difficult degradation. If being directly discharged without being treated, it will bring serious harm to the ecological environment. Because of the dangers of dyeing wastewater, many countries have enacted strict emissions standards, but There is no uniform standard currently.

Waste minimization is of great importance in decreasing pollution load and production costs. This book has shown that various methods can be applied to treat cotton textile effluents and to minimize pollution load. Traditional technologies to treat textile wastewater include various combinations of biological, physical, and chemical methods, but these methods require high capital and operating costs. Technologies based on membrane systems are among the best alternative methods that can be adopted for large-scale ecologically friendly treatment processes. A combination methods involving adsorption followed by nanofiltration has also been advocated, although a major drawback in direct nanofiltration is a substantial reduction in pollutants, which causes permeation through flux.

It appears that an ideal treatment process for satisfactory recycling and reuse of textile effluent water should involve the following steps. Initially, refractory organic compounds and dyes may be electrochemically oxidized to biodegradable constituents before the wastewater is subjected to biological treatment under aerobic conditions. Color and odor removal may be accomplished by a second electrooxidation process. Microbial life, if any, may be destroyed by a photochemical treatment. The treated water at this stage may be used for rinsing and washing purposes; however, an ion-exchange step may be introduced if the water is desired to be used for industrial processing.

As the improvement of the environmental protection laws and the raise of the awareness of environmental protection, the pollution of printing and dyeing enterprises has caught a lot of attention and the treatment of dyeing wastewater has become a focus. Recently, except the oxidation, filtration and other single method research, it has been introduced a large number of electric, magnetic, optical and thermal method to treat the refractory materials. Several of other techniques have also been carried out to treat the wastewater in order to develop a variety of ways. The interdisciplinary study based on the traditional biological approach will be the direction of the wastewater treatment. The writer expected that the technology which is efficient, clean and reasonable will come out soon.

On the other hand, clean production is also an important research, which can shift the focus from end of the treatment to the prevention of pollution and conduct more in-depth research on the printing and dyeing production technology and process management. Moreover, the strategic, comprehensive, preventive measures and advanced production technology can be used to improve the material and energy utilization. Also, we can reduce and eliminate the generation and emissions of wastes as well as the production of excessive use of resources and the risks to humans and the environment.

Prevention and treatment of dyeing wastewater pollution are complementary. We can both use preventive measures as well as a variety of methods to control the wastes and make use of treated water. This will not only reduce water consumption, but also effectively reduce the pollution of the printing and dyeing wastewater and achieve sustainable development of society.

4. References

B. Ramesh Babu; A.K. Parande; S. Raghu; and T. Prem Kumar (2007). Textile Technology-Cotton Textile Processing: Waste Generation and Effluent Treatment. *The Journal of Cotton Science* 11(2007) 141–153

C. All`egre; P. Moulin; M. Maisseu; F. Charbit (2006) .Treatment and reuse of reactive dyeing effluents. *Journal of Membrane Science* 269 (2006) 15-34

Compilers. "Discharge standard of water pollutants for dyeing and finishing of textile industry". *State bureau of environmental protection* 4(2008) (in Chinese)

Jia Haidong (2003). Application of SBR in Dyeing Wastewater Treatment. *Environmental Review* 16(2003) (in Chinese)

Joseph Egli Italia srl (2007). Wastewater treatment in the textile industry. *Dyeing Printing Finishing* 10(2007)60-66

K. Ranganathan; K. Karunagaran; D.C. Sharma (2007).Recycling of wastewaters of textile dyeing industries using advanced treatment technology and cost analysis — Case studies. *Resources, Conservation and Recycling* 50 (2007) 306–318

Li-yan Fu; Xiang-hua Wen *; Qiu-li Lu; Yi Qian (2001). Treatment of dyeing wastewater in two SBR systems. *Process Biochemistry* 36 (2001) 1111–1118

Li Honglian (2006). The dyeing wastewater treatment process of hydrolysis - biological contact Oxidation – Aeration. *Industrial Water and Wastewater* 11(2006) (in Chinese)

Mattioli D.; Malpei F.; Bortone G.;Rozzi A. (2002). Water minimization and reuse in the textile industry. *Water Recycling and resource recovery in industry* 4(2002)

State Environmental Protection Administration (1994), Textile Association, Eighth Five-Year Research Project Team. Report of the textile industrial pollution control. *China Environmental Science Press,*1994 (in Chinese)

Sheng H.Lin,Ming L.Chen (1997). *Treatment of textile wastewater by chemical methods for reuse.* Wat.Res 31(1997)868-876.

Sheng-Jie You; Dyi-Hwa Tseng; Jun-Yu Deng (2008). Using combined membrane processes for textile dyeing wastewater reclamation. *Desalination* 234 (2008) 426–432

Shaolan Ding; Zhengkun Li; Wangrui (2010). Overview of dyeing wastewater treatment technology. *Water resources protection* 26(2010) 73-78 (in Chinese)

Wu Zhimin; Zhangli (2009). Coagulation-ABR-Oxidation ditch process of dyeing wastewater. *Printing and Dyeing* 3(2009) (in Chinese)

Yan Kelu (2005). "A Course in Dyeing and Finishing". Chinese ming press, 2005 (in Chinese)

Zeng Kangmei; Li Zhengshan; Wei Wenwen (2005). Industrial production and pollution control. *Chemical Industry Press* 4(2005)101-104 (in Chinese)

Zhang Kerong (2007). UASB-aerobic Physico -chemical treatment of dyeing wastewater. *Resources and Environment,* 2007 (in Chinese)

3

Photochemical Treatments of Textile Industries Wastewater

Falah Hassan Hussein

Chemistry Department, College of Science, Babylon University
Iraq

1. Introduction

Textile industry is one of the most water and chemical intensive industries worldwide due to the fact that 200-400 liters of water are needed to produce 1 kg of textile fabric in textile factories (Correia et al., 1994 & Orhon et al., 2003). The water used in this industry is almost entirely discharged as waste. Moreover, the loss of dye in the effluents of textile industry can reach up to 75% (Couto & Toca-Herrera, 2006). The effluents are considered very complex since they contain salt, surfactants, ionic metals and their metal complexes, toxic organic chemicals, biocides and toxic anions.

Azo dyes are regarded as the largest class of synthetic. Approximately, 50–70% of the available dyes for commercial applications are azo dyes followed by the anthraquinone group (Konstantinou & Albanis, 2004). Azo dyes are classified according to the presence of azo bonds (–N=N–) in the molecule i.e., monoazo, diazo , triazo etc. and also sub-classified according to the structure and method of applications such as acid, basic, direct, disperse, azoic and pigments (Bhutani, 2008). Some azo dyes and their dye precursors are well-known of high toxicity and suspected to be human carcinogens as they form toxic aromatic amines (Gomes et al., 2003 & Stylidi et al., 2003).

Different physical, chemical and biological as well as the various combinations of pre-treatment and post-treatment techniques have been developed over the last two decades for industrial wastewaters treatment in order to meet the ever-increasing requirements of human beings for water. Though there are numerous studies published in this field, most of the techniques adopted by these researchers are uneconomical, ineffective or impractical uses (Cooper, 1995 & Stephen, 1995).

Recent studies have demonstrated that heterogeneous photocatalysis is the most efficient technique in the degradation of colored chemicals(Li et al., 2003; Vione et al., 2003; Antharjanam et al., 2003; Fernandez-Ibanez et al., 2003; Liu et al., 2003; Ohno, 2004; Chen et al., 2004; Alkhateeb et al., 2005 & Attia et al., 2008). These studies used titanium dioxide and / or zinc oxide in the photolysis processes. The large bang gap of titanium dioxide and zinc oxide (~ 3.2 eV) put a limitation of using these semiconductors in photocatalytic degradation under natural weathering conditions. Only a small part of the overall solar intensity could be useful in such photodegradation processes. However, the existence of dye on the surface of catalyst reduces the energy required for excitation and then increases the efficiency of the excitation process by extending its absorption in the visible region of the spectrum.

2. Methods used in treatment of textile wastewater

There are several factors to choose the appropriate textile wastewater treatment method such as, economic efficiency, treatment efficiency, type of dye, concentration of dye and environmental fate.

There is no general method for the treatment of textile industrial wastewater. Wastewaters from textile industry contain various pollutants resulting from various stages of production, such as, fibers preparation, yarn, thread, webbing, dyeing and finishing. Mainly three methods are used for the treatment of textile industrial wastewater. These are:

1.   Physico-chemical methods.
2.   Advanced oxidation methods.
3.   Biological sludge methods.

The main operational methods used for the treatment of textile industrial water involve physical and chemical processes (Shaw et al., 2002 & Liu et al., 2005). However, these techniques have many disadvantages (See Table 1, Mutambanengwe, 2006). These disadvantages include Sludge generation, high cost , formation of bi – products, releasing of toxic molecules, requiring a lot of dissolved oxygen, limitation of activity for specific dyes and requiring of long time.

In recent years, advanced oxidation processes (AOPs) have gained more attraction as a powerful technique in photocatalytic degradation of textile industrial wastewater since they are able to deal with the problem of dye destruction in aqueous systems (Konstantinou & Albanis, 2004).

3. Photocatalysis

Photocatalysis is defined as the acceleration of a photoreaction in the presence of a catalyst, while photolysis is defined as a chemical reaction in which a chemical compound is broken down by photons. In catalyzed photolysis, light is absorbed by an adsorbed substrate.

Photocatalysis on semiconducting oxides relies on the absorption of photons with energy equal to or greater than the band gap of the oxide, so that electrons are promoted from the valence band to the conduction band:

$$Semiconductor + hv \rightarrow h^+ + e^- \tag{1}$$

If the photoholes and photoelectrons produced by this process migrate to the surface, they may interact with adsorbed species in the elementary steps, which collectively constitute photocatalysis.

The numerous gas- and liquid-phase reactions photocatalysed by TiO_2 and ZnO have been reviewed. There is a general agreement that adsorption is necessary since the surface species act as traps for both; photogenerated holes and electrons, which otherwise recombine.

In addition to participating in conventional surface reaction steps, adsorbed dye molecules assist in the separation of photoholes and photoelectrons, which may otherwise recombine within the semiconductor particles. A major factor affecting the efficiency of photocatalysis process is electron/hole recombination.

If the electrons and holes are used in a reaction, a steady state will be reached when the removal of electrons and holes equals the rate of generation by illumination. Recombination and trapping processes are the de-excitation processes which are responsible for the creation of the steady state, if no reaction occurs. There are three important mechanisms of recombination:

1. Direct recombination.
2. Recombination at recombination centers.
3. Surface recombination.

There are different types of semiconductors, whose band gaps range between 1.4 and 3.9 eV, i.e., it could be excited with a light of 318–886 nm wavelengths. This means that most of the known semiconductors could be excited by using visible light. However, not all these semiconductors could be used in the photocatalytic reactions. Bahnemann et al., (1994) report that the most appropriate photocatalysts should be stable toward chemicals and illumination and devoid of any toxic constituents, especially for those used in environmental studies. The authors also explaine that TiO_2 and ZnO are the most commonly used in photocatalytic reactions due to their efficient absorbtion of long wavelength radiation as well as their stability towards chemicals. Other semiconductors like WO_3, CdS, GaP, CdSe and GaAs absorb a wide range of the solar spectrum and can form chemically activated surface-bound intermediates, but unfortunately these photocatalysts are degraded during the repeated catalytic cycles involved in heterogeneous photocatalysis.

4. Photosensitization

The illumination of suspended semiconductor in an aqueous solution of dye with unfiltered light (polychromatic light) leads to the possibility of the existence of two pathways (Hussein et al., 2008):

1. In the first pathway, the part of light with energy equal to or more than the band gap of the illuminated semiconductor will cause a promotion of an electron to conduction band of the semiconductor and as a result, a positive hole will be created in the valence band. The formed photoholes and photoelectrons can move to the surface of the semiconductor in the presence of light energy. The positive hole will react with adsorbed water molecules on the surface of semiconductor producing $^\bullet OH$ radicals and the electron will react with adsorbed oxygen on the surface. Moreover, they can react with deliquescent oxygen and water in suspended liquid and produce perhydroxyl radicals ($HO_2^\bullet$) with high chemical activity (Zhao & Zhang, 2008). The processes in this pathway can be summarized by the following equations:

$$h^+ + OH^- \rightarrow \overset{\bullet}{O}H \tag{2}$$

$$h^+ + H_2O \rightarrow H^+ + \overset{\bullet}{O}H \tag{3}$$

$$e^- + O_2 \rightarrow O_2^{\bullet -} \tag{4}$$

$$O_2^{\bullet -} + H^+ \rightarrow HO_2^{\bullet} \tag{5}$$

$$HO_2^{\bullet} + O_2^{\bullet -} \xrightarrow{\ H^+\ } H_2O_2 + O_2 \tag{6}$$

$$H_2O_2 \rightarrow 2\overset{\bullet}{O}H \tag{7}$$

$$\text{Dye} + \text{Semiconductor}\,(h^+{}_{VB}) \rightarrow \text{Dye}^{\cdot\,+} + \text{Semiconductor} \tag{8}$$

$$\text{Dye}^{\cdot\,+} + O_2^{\cdot\,-} \rightarrow DO_2 \rightarrow \text{degradation products} \tag{9}$$

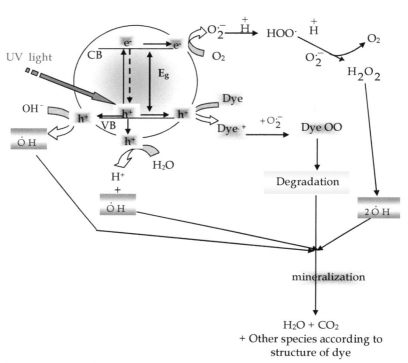

Fig. 1. Dye/semiconductor/UV light system

2.   In the second pathway, the other part of light with energy which is less than the band gap of the illuminated semiconductor will be absorbed by the adsorbed dye molecules. Dye molecules will be decolorized by a photosensitization process. The photocatalytic decolorization of dyes, which is described as photosensitization processes, is also characterized by a free radical mechanism. In this process, the adsorbed dyes molecules on the surface of the semiconductor can absorb a radiation in the visible range in addition to the radiation with a short wavelengths (Fernandez-Ibanez et al., 2003; Ohno, 2004 & Alkhateeb et al., 2005). The excited colored dye (dye*) (in the singlet or triplet state) will inject an electron to the conduction band of the semiconductor (Hussein & Alkhateeb, 2007). The processes in this pathway can be summarized by the following equations:

$$\text{Dye} + h\upsilon\,(\text{VIS or UV reagion}) \rightarrow {}^1\text{Dye}^* \text{ or } {}^3\text{Dye}^* \tag{10}$$

$${}^1\text{Dye}^* \text{or } {}^3\text{Dye}^* + \text{Semiconductor} \rightarrow \text{Dye}^{\cdot\,+} + e^- \text{(to the conduction band of Semiconductor)} \tag{11}$$

$$e^- + O_2 \rightarrow O_2^{\cdot -} + Semiconductor \tag{12}$$

$$Dye^{\cdot +} + O_2^{\cdot -} \rightarrow DyeO_2 \rightarrow degradation\ products \tag{13}$$

$$Dye^{\cdot +} + HO_2^{\cdot}(or\ H\dot{O}) \rightarrow degradation\ products \tag{14}$$

$$Dye + 2\dot{O}H \rightarrow H_2O + oxidation\ products \tag{15}$$

$$Dye^{\cdot +} + \overline{O}H \rightarrow Dye + \dot{O}H \tag{16}$$

$$Dye^{\cdot +} + H_2O \rightarrow Dye + \dot{O}H + H^+ \tag{17}$$

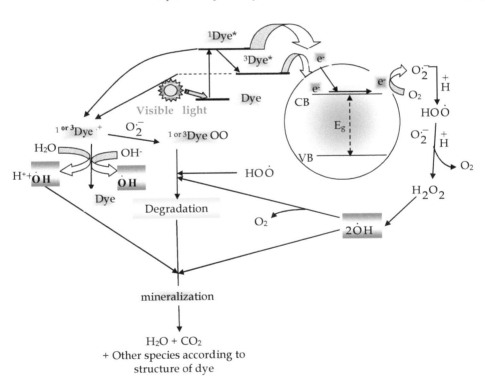

Fig. 2. Nonregenerative dye/semiconductor/visible light system

The mechanism above is favoured by nonregenerative organic dye where dye/semiconductor/visible light system and the sensitizer itself degrade. However, in regenerative semiconductor system, the following mechanism may be followed:

$$Sen + TiO_2 \leftrightarrow Sen - TiO_2 \tag{18}$$

$$Sen - TiO_2 + RX \leftrightarrow Sen - TiO_2......RX_{(ads)} \tag{19}$$

$$Sen - TiO_2 + O_2 \leftrightarrow Sen - TiO_2......O_{2(ads)} \tag{20}$$

$$Sen - TiO_2 \xrightarrow[h\upsilon]{h\upsilon \; visible\,light} Sen^* - TiO_2 \tag{21}$$

$$Sen^* - TiO_2......O_{2(ads)} \rightarrow Sen^+ - TiO_2 + O_2^- \tag{22}$$

$$Sen^+ - TiO_2 + O_2^- \rightarrow Sen - TiO_2 + O_2 \tag{23}$$

$$Sen^* - TiO_2 \rightarrow Sen^+ - TiO_2\left(e_{C.B}^-\right) \tag{24}$$

$$Sen^+ - TiO_2\left(e_{C.B}^-\right) \rightarrow Sen - TiO_2 \tag{25}$$

$$Sen^+ - TiO_2\left(e_{C.B}^-\right)......RX \rightarrow Sen^+ - TiO_2 + \dot{R}X + X^- \tag{26}$$

$$Sen^+ - TiO_2\left(e_{C.B}^-\right)......O_{2(ads)} \rightarrow Sen^+ - TiO_2 + O_2^- \tag{27}$$

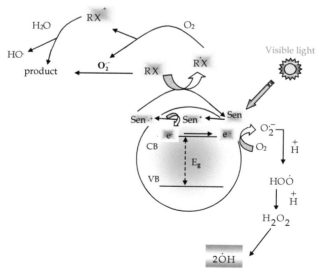

Fig. 3. Regenerative dye/semiconductor/visible light system

Cho et al., (2001) conclude that there is no direct electron transfer between an excited sensitizer and CCl_4 molecules, in homogeneous solution, and the existence of semiconductor is essential for sensitized photocatalysis. Platinum supported on titanium dioxide acts as an excellent sensitizer and could have practical advantages as a mild and convenient photocatalyst for selective oxidation processes (Hussein et al., 1984). The addition of Rhodamine-B, as sensitizer to TiO_2 dispersion system increases the rate of photooxidation properties (Hussein et al., 1991). The authors explained that due to the fact that more light absorbed by Rhodamine-B between 460-580 nm, then the energy transfer from sensitizer to TiO_2 or to any other active species and hence promote the photocatalytic activity of titanium dioxide.

5. Advanced oxidation processes

Glaze et al., (1987) define Advanced Oxidation Processes (AOPs) for water treatments as the processes that occur near ambient temperature and pressure which involve the generation of highly reactive radicals ,especially hydroxyl radicals ($\bullet$OH), in sufficient quantity for water purification. Advanced oxidation processes can also be easily defined as techniques of destruction of organic pollutants from wastewaters. These processes include chemical oxidation processes using hydrogen peroxide, ozone, combined ozone and hydrogen peroxide, hypochlorite, Fenton's reagent, ultra-violet enhanced oxidation such as UV/O_3, UV/ H_2O_2, UV/air, wet air oxidation and catalytic wet air.

Hydroxyl radicals are strong reactive species, which are capable of destroying a wide range of organic pollutants. Table 1 shows hydroxyl radical as the second strongest oxidant (Weast, 1977; Legrini et al., 1993 Domènech et al., 2001 ; & Mota et al., 2008).

Oxidant	E° (V)
Fluorine (F_2)	3.03
Hydroxyl radical ($\bullet$OH)	2.80
Atomic oxygen (O)	2.42
Ozone (O_3)	2.07
Hydrogen peroxide(H_2O_2)	1.78
Hydroperoxyl radical ($O_2H \cdot$)	1.70
Potassium permanganate ($KMnO_4$)	1.67
Hypobromous acid (HBrO)	1.59
Chlorine dioxide (ClO_2)	1.50
Hypochlorous acid (HClO)	1.49
Hypochloric acid	1.45
Chlorine (Cl_2)	1.36
Bromine (Br_2)	1.09
Iodine (I_2)	0.54

Table 1. Standard reduction potential of common oxidants against Standard Hydrogen Electrode

The attack of organic pollutants by hydroxyl radicals occurs via the following mechanisms (Buxton et al., 1988; Legrini et al., 1993 & Pignatello et al., 2006):

1. Electron transfer from organic pollutants to hydroxyl radicals:

$$\dot{O}H + RX \rightarrow RX^{\cdot +} + OH^- \tag{28}$$

2. Hydrogen atom abstraction from the C-H, N-H or O-H bonds of organic pollutants:

$$\dot{O}H + RH \rightarrow R^{\cdot} + H_2O \tag{29}$$

3. Addition of hydroxyl radical to one atom of a multiple atom compound:

$$\dot{O}H + Ph \rightarrow HOPh^{\cdot} \tag{30}$$

6. Fundamental parameters in photocatalysis

In semiconductor photocatalysis of industrial wastewater treatment, there are different parameters affecting the efficiency of treatment. These parameters include mass of catalyst, dye concentration, pH, light intensity, addition of oxidizing agent, temperature and type of photocatalyst. Other factors, such as, ionic components in water, solvent types, mode of catalyst application and calcinations temperature can also play an important role on the photocatalytic degradation of organic compounds in water environment (Guillard et al., 2005 & Ahmed et al., 2011).

6.1 Effect of type of catalyst

Haque & Muneer (2007) observe that Degussa P25 is more reactive for degradation of a textile dye derivative, bromothymol blue, in aqueous suspensions than other commercially available photocatalysts types of titanium dioxide, namely Hombikat UV100, PC500 and TTP. They explain the high activity of Degussa P25 is due to composing of small nano-crystallites of rutile dispersed within the anatase matrix. The band gap of rutile is less than that of anatase and as a result electron will transfer from the rutile conduction band to electron traps in anatase and the recombination of electrons and holes will be reduced. Hussein, (2002) reported that anatase has higher photoactivity than rutile due to the difference in surface area.

Decolorization percentage of real textile industrial wastewater on rutile, anatase, and zinc oxide shows that the activity of different catalysts falls in the following sequence (Hussein & Abass, 2010 a):

$$ZnO > TiO_2 \text{ (Anatase)} > TiO_2 \text{ (Rutile)}$$

ZnO is more active than TiO_2 due to the absorption of wider spectrum light (Sakthivela et al., 2003). However, the amount of zinc oxide required to reach the optimum activity is two times more than that for titanium dioxide (anatase or rutile) (Hussein & Abass, 2010 a). In another study, Hussein et al., (2008) observed that ZnO is less active than anatase when the same weight of catalysts is used for photocatalytic degradation of textile wastewater. Akyol et al., (2004) reported that ZnO is more active than TiO_2 for the decolorization efficiency of aqueous solution of a commercial textile dye due to the band gap energy, the charge carrier density, and the crystal structure.

Decolorization efficiency of real textile industrial wastewater in the presence and absence of catalyst and/or solar radiation was also investigated (Hussein & Abass, 2010 b). The results indicate that the activity of different catalysts fall in the sequence:

ZnO > TiO$_2$ (Anatase) > TiO$_2$ (Rutile) > in the absence of catalyst = in the absence of solar radiation or artificial radiation = 0

The results are plotted in Figure 4.

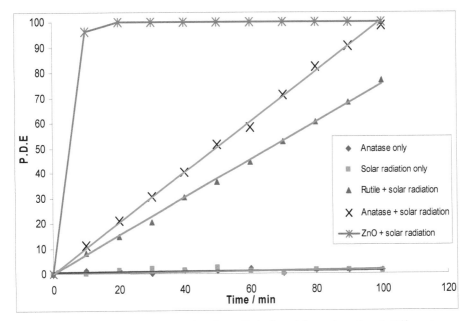

Fig. 4. Photocatalytic decolorization of real textile industrial wastewater at different conditions.

These results also indicate that there has been no dark reaction. Incubations of colored industrial wastewater without solar radiation and/or without catalyst has been performed to demonstrate that decolorization of the dye is dependent on the presence of both; light and catalyst.

6.2 Effect of mass of catalyst

Photocatalysts dosage added to the reaction vessel is a major parameter affecting the photocatalytic degradation efficiency (Dong et al., 2010). Photocatalytic degradation efficiency increases with an increase in catalysts mass. This behavior may be due to an increase in the amount of active site on surface of photocatalyst particles. As a result, an increasing the number of dye molecule adsorbed on the surface of photocatalyst lead to an increase in the density of particles in the area of illumination (Kim & Lee, 2010). The extrapolation of Hird's data (Hird, 1976) indicates that only 7.5 mg of TiO$_2$ was sufficient to absorb all incident 366 nm radiation. It follows that the mass effect must be caused by changes in the effective utilization of the absorbed radiation rather than by the increased absorption.

Photocatalyst with small particles are more efficient than larger particles. This behavior may be due to (Hussein, 1984):

1.   Photoholes and photoelectrons generated in the bulk would have fewer traps and recombination centers to overcome before reaching the surface.
2.   A greater proportion of material would be within the space charge arising from depletive oxygen chemisorptions, which favor exciton dissociation and photohole migration to the surface.

Hence, increasing the catalyst's mass will increase the concentration of the efficient small particles within the illuminated region of the reaction vessel. The direct proportionality between photocatalytic degradation efficiency and catalyst loading is real within low concentrations of photocatalyst where there are excess active sites reaching plateau reign. The plateau is reached when this effect can no longer increase the overall efficiency of utilizing incident radiation. Moreover, after the plateau region is achieved, the activity of photocatalytic decolorization decrease with increase of catalyst concentration for all types of catalysts. This behavior is more likely to emanate from variation in the intensity of radiation entering the reaction vessel and the way the catalyst utilizes that radiation. Light scattering by catalyst particles at higher concentration lead to decrease in the passage of irradiation through the sample leading to poor light utilization (Gaya et al., 2010; Kavitha & Palanisamy, 2011). Deactivation of activated photocatalyst molecules colliding ground state molecules with increasing the load of photocatalyst may be also cause reduction in photocatalyst activity (Kim & Lee, 2010).

Photocatalytic decolorization efficiency (PDE) % of real textile industrial wastewater has been investigated by employing different masses of TiO_2 (anatase or rutile) or ZnO under natural weathering conditions for 20 minutes of irradiation (Hussein & Abass, 2010 b). The results are plotted in Figure 5.

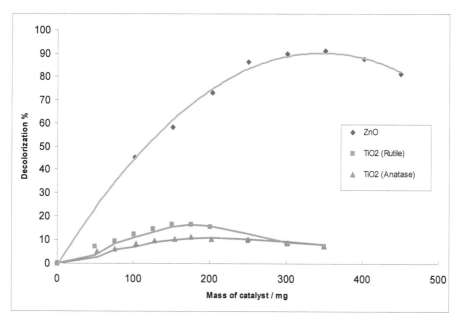

Fig. 5. Effect of mass on photocatalytic decolorization efficiency of real textile industrial wastewater

The results in all cases indicate that the decolorization efficiency increases with increase in catalysts mass and then it becomes constant. It is clear from consideration of the catalyst concentrations at which the activity plateau were achieved that the mass effect does not depend upon the type of dye and source of irradiation. Moreover, plateau regions were achieved and then the activity of decolorization decreased with increasing catalyst concentration, for all types of catalysts used in this project.

6.3 Effect of pH
Aqueous solution pH is an important variable in the evaluation of aqueous phase mediated photocatalytic decolorization reactions. pH change effects the adsorption quantity of organic pollutants and the ways of adsorption on the surface of photocatalyst (coordination). As a result, the photocatalytic degradation efficiency will greatly be influenced by pH changes.

Zero Point Charge (pHzpc), is a concept relating to adsorption phenomenon and defined as the pH at which the surface of an oxide is uncharged. If positive and negative charges are both present in equal amounts, then this is the isoelectric point (iep). However, the zpc is the same as iep when there is no adsorption of other ions than the potential determining H^+/OH^- at the surface.

In aqueous solution, at pH higher than pHzpc, the oxide surface is negatively charged and then the adsorption of cations is favoured and as a consequence, oxidation of cationic electron donors and acceptors are favoured. At pH lower than pHzpc, the adsorbent surface is positively charged (See Figure 6) and then the adsorption of anions is favoured and as a consequence, the acidic water donates more protons than hydroxide groups.

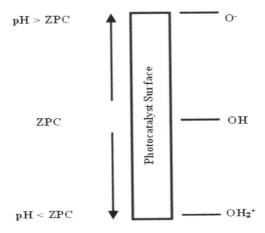

Fig. 6. Effect of pH on ZPC.

Infrared spectroscopy study of Szczepantiewicz et al., (2000) shows that the TiOH sites are the major electron traps when TiO_2 is illuminated. The distribution of other species ($TiOH_2^+$ and TiO^-) with changing pH has been proposed by Kormann & co-workers (1991) (See Figure 7). Figures 6 and 7 show that , at pH below ZPC the surface is mostly positively charged and TiOH sites increase as pH increases and reach maximum value at ZPC of semiconductor. However, $TiOH_2^+$ as pH increases and reaches zero value at ZPC. At pH higher than ZPC the density of TiO⁻ groups on the surface start to form and reached 100% value at pH 14. The importance of pH during the reaction is not less than that of initial state. The formation of intermediate products, sometimes, changes the pH of aqueous solution and as a result, it affects the rate of photodegeradation (Galvez, 2003).

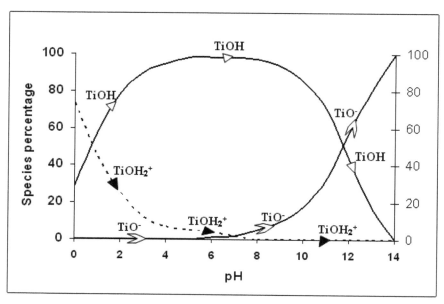

Fig. 7. The distribution of TiO₂ surface species at different pH

At pH above and below pHzpc, the surface of zinc oxide and titanium dioxide are negatively or positively charged according to the following equations:

$$ZnOH + H^+ \leftrightarrow ZnOH_2^+ \tag{31}$$

$$ZnOH + OH^- \leftrightarrow ZnO^- + H_2O \tag{32}$$

$$TiOH + H^+ \leftrightarrow TiOH_2^+ \tag{33}$$

$$TiOH + OH^- \leftrightarrow TiO^- + H_2O \tag{34}$$

6.3.1 Effect of pH on photocatalytic decolorization of Bismarck brown R
Under the determined experimental condition with initial dye concentration equal to 10^{-4} M, ZnO dosage 3.75 gm.L⁻¹, light intensity equal to 2.93 mW.cm⁻² and temperature equal to 298.15 K, the effect of change in solution pH on decolorization percentage has been studied in the range 2-12 (Figure 8). The decolorization percent has been found to be strongly dependent on pH of solution because the reaction takes place on the surface of semiconductor. The decolorization percentage of Bismarck brown R increases with the increase of pH, exhibiting maximum decolorization at pH 9.

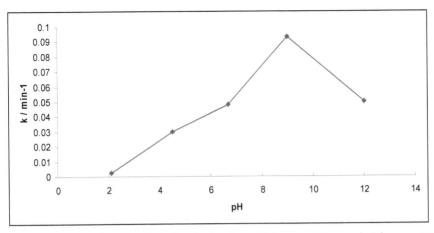

Fig. 8. Effect of pH on photocatalytic decolorization efficiency of Bismarck R brown on ZnO

Under the determined experimental condition with initial dye concentration equal to 10^{-4}M, TiO_2 dosage 1.75 gm.L^{-1}, light intensity equal to 2.93 mW.cm^{-2} and temperature equal to 298.15 K, the effect of change in solution pH on decolorization percentage has been studied in the range 2-10. The results are plotted in Figures 9, for TiO_2 (DegussaP25), TiO_2 (HombikatUV100), TiO_2 (MillenniumPC105) and TiO_2 (Koronose2073). It was observed that the decolorization percentage strongly depends on the pH of solution because the reaction takes place on the surface of semiconductor. The decolorization percentage of Bismarck brown R increases with the increase of pH, exhibiting maximum decolorization at pH that is equal to 6.61, 6.54, 6.75, 6.63 for TiO_2 (DegussaP25), TiO_2 (HombikatUV100), TiO_2 (MillenniumPC105) and TiO_2 (Koronose2073), respectively.

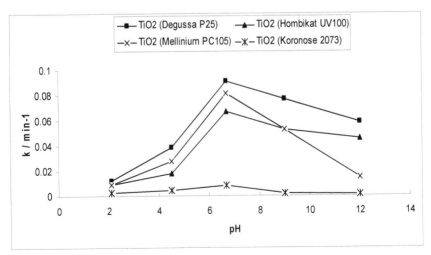

Fig. 9. Effect of pH on photocatalytic decolorization efficiency of Bismarck R brown on different types of TiO_2

This behavior could be explained (as mentioned before) on the basis of zero point charge (ZPC). The zero point charge is equal to 9.00 for ZnO and 6.25 for TiO$_2$ (Degussa P25). With the increase of the pH of solution, the surface of catalyst will be negatively charged by adsorbed hydroxyl ions. The presence of large quantities of adsorbed OH$^-$ ions on the surface of catalyst favor the formation of $\cdot$OH radical. However, if pH is lower than ZPC, the hydroxyl ions adsorbed on the surface will be decreased and, therefore, hydrogen ions adsorbed on the surface will increase and the surface will become positive charged. Both the acidic and basic media leave an inverse impact on the photodecolorization efficiency because of the decrease of the formation of the hydroxyl radical.

The decolorization of Bismarck brown R decreases dramatically at strong acid media (pH = 2.1) for ZnO. This could be explained due to photocorrosion of ZnO as shown in the following equations:

$$ZnO \xrightarrow{\ h\upsilon\ } e^-_{CB} + h^+_{VB} \tag{35}$$

$$ZnO + 2h^+_{VB} \rightarrow Zn^{2+} + \frac{1}{2}O_2 \tag{36}$$

6.4 Effect of Light Intensity

Egerton & King, (1979) show that square root of light intensity depends on the activity of titanium dioxide for different wavelengths of light. However, this relationship cannot be applied to all range of light intensities. The primary electronic processes which occur in the absorption of photons with energy equal or greater than the band gap of semiconductor are:

$$\text{Semiconductor} + h\upsilon \xrightarrow{\ k_1\ } (h-e)\ \text{exciton} \tag{37}$$

$$(h-e) \xrightarrow{\ k_2\ } h^+ + e^- \tag{38}$$

$$h^+ + e^- \xrightarrow{\ k_3\ } \text{Radiationless Recombination} \tag{39}$$

For photocatalysis processes, it is necessary that the excitons dissociate and the photoholes and photoelectrons reach the catalyst surface where they are trapped by surface species:

$$h^+ + OH^-_{(s)} \xrightarrow{\ k_4\ } \dot{O}H_{(s)} \tag{40}$$

$$e^- + O_2 \xrightarrow{\ k_5\ } O^-_{2(ads)} \tag{41}$$

The concentration of excitons, photoholes and photoelectrons may be considered by applying a steady state treatment:

$$\frac{d[h-e]}{dt} = k_1 I_{(abs)} - k_2 \big[(h-e)\big] = 0 \tag{42}$$

So that:

$$[(h-e)] = \frac{k_1}{k_2} I_{(abs)} \tag{43}$$

Similarly:

$$\frac{d[h]}{dt} = k_2[(h-e)] - k_3[h^+][e^-] - k_4[h^+]\left[OH_{(S)}^-\right] = 0 \tag{44}$$

Since:

$$[h^+] = [e^-] \tag{45}$$

Then:

$$\frac{d[h]}{dt} = k_1 I_{(abs)} - k_3[h^+]^2 - k_4[h^+]\left[OH_{(s)}^-\right] = 0 \tag{46}$$

So that:

$$k_1 I_{(abs)} = k_3[h^+]^2 + k_4[h^+]\left[OH_{(s)}^-\right] \tag{47}$$

There are two possibilities concerning the light intensity:
a.   At high light intensities, where the recombination of photoholes and photoelectrons is predominate, then:

$$k_3[h^+]^2 \gg k_4[h^+]\left[OH_{(s)}^-\right] \tag{48}$$

So equation 48 becomes:

$$k_1 I_{(abs)} = k_3[h^+]^2 \tag{49}$$

Then:

$$[h^+] = \left(\frac{k_1}{k_3}\right)^{\frac{1}{2}} I_{(abs)}^{\frac{1}{2}} \tag{50}$$

If the rate controlling step in the overall photocatalysis processes involves the surface trapping of photoholes at surface OH-, then the rate will be given by equation 40.
The reaction rate is given by:

$$\text{Reaction rate} = k_4 \left(\frac{k_1}{k_3}\right)^{\frac{1}{2}} I_{(abs)}^{\frac{1}{2}}\left[OH_{(s)}^-\right] \tag{51}$$

Then:

$$\text{Reaction rate } \alpha \left[I_{(abs)} \right]^{\frac{1}{2}} \tag{52}$$

If, on the other hand, the rate controlling step involves photoelectron trapping by oxygen, then the rate controlling step will be equation 41. The reaction rate is given by:

$$\text{Reaction rate} = k_5 \left(\frac{k_1}{k_3} \right)^{\frac{1}{2}} I_{abs} \left[O_{2(ads)} \right] \tag{53}$$

Then:

$$\text{Reaction rate } \alpha \left[I_{(abs)} \right]^{\frac{1}{2}} \tag{54}$$

b.   At low light intensities, it is expected that recombination of photoholes and photoelectrons will be low, then:

$$k_4 \left[h^+ \right] \left[OH^-_{(s)} \right] \gg k_3 \left[h^+ \right]^2 \tag{55}$$

So equation 47 becomes:

$$k_1 I_{(abs)} = k_4 \left[h^+ \right] \left[OH^-_{(s)} \right] \tag{56}$$

Then:

$$\left[h^+ \right] = \frac{k_1 \left[I_{(abs)} \right]}{k_4 \left[OH^-_{(s)} \right]} \tag{57}$$

It follows that:

$$\text{Reaction rate} = k_1 \left[I_{(abs)} \right] \tag{58}$$

Alternatively, if photoelectron trapping is considered to be rate controlling, then:

$$\text{Reaction rate} = k_5 \frac{k_1}{k_4} \left[\left(I_{(abs)} \right) \right] \tag{59}$$

Hence, a linear dependence would be expected at low light intensities. Square-root intensity dependence was observed with rutile I, rutile 11, anatase, uncoated anatase pigment and platinized anatase and also independent on wavelength of incident radiation (Harvey et al., 1983 a; Hussein & Rudham, 1987). Bahnemann et al., (1991) reported that the change in kinetic constant is a function of the square root of the radiation entering at high light intensities, while this change can be linear with light intensity of incident radiation low light intensities (Peterson et al., 1991).

Ollis et al., (1991) summarize the effect of light intensity on the kinetics of the photocatalytic degradation of dye as follows:

a.   At low light intensities (0–20 mW/cm²), the rate of photocatalytic degradation is proportional directly with light intensity (first order).

b.   At high light intensities (25 mW/cm²), the rate of photocatalytic degradation is proportional directly with the square root of the light intensity (half order).

c.   At high light intensities the rate of photocatalytic degradation is independent of light intensity (zero order). See Figure 10.

However, Hussein et al (2011) found that the rate of photocatalytic decolorization of Bismarck brown R on ZnO and different types of titanium dioxide is proportional directly with the light intensity of incident UVA radiation in the range of 0- 2.0 mW/cm² and with the square root of light intensity in the range of 2.0- 3.5 mW/cm².

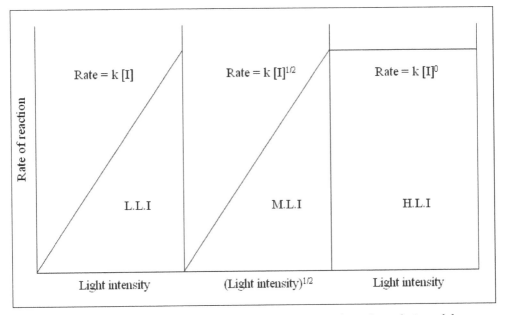

Fig. 10. Effect of light intensity on the kinetics of the photocatalytic degradation of dye

Figures 11 and 12 illustrate the impact of initial light intensity on the value of rate constant for photocatalytic decolorization of Bismarck brown R on ZnO and different types of titanium dioxide, respectively. The results indicate that the photocatalytic decolorization of Bismarck brown R increases with the increase in light intensity, attaining a maximum value at 3.52 mW.cm-².

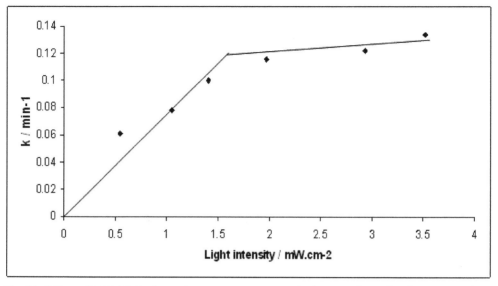

Fig. 11. Effect of initial light intensity on rate constant of photocatalytic decolorization of Bismarck brown R on ZnO

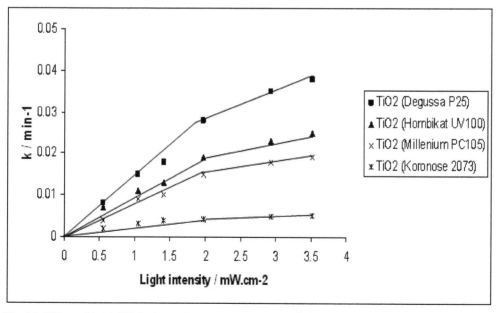

Fig. 12. Effect of initial light intensity on rate constant of photocatalytic decolorization of Bismarck brown R using different types of TiO_2

6.5 Effect of temperature

One of the advantages of photoreaction is that it is not affected or slightly affected by temperature change. Temperature dependent steps in photocatalytic reaction are adsorption and desorption of reactants and products on the surface of photocatalyst. None of these steps appears to be rate determining. The impact of temperature is explained as the variable with the least effect on photocatalytic degradation of aqueous solution of azo dyes (Obies, 2011). Attia et al., (2008) have found that the activation energy of photodegeradation of real textile industrial wastewater is equal to 21 ± 1 kJ mol[-1] on titanium dioxide and 24 ± 1 kJ mol[-1] on zinc oxide. The activation energy for the photocatalytic degradation of textile industrial wastewater on titanium dioxide is similar to previous findings for photocatalytic oxidation of different types of alcohols on titanium dioxide and metalized titanium dioxide (Al-zahra et al., 2007; Hussein & Rudham, 1984, 1987). The single value of activation energy (21 ± 1 kJ mol[-1]) that can be related to the calculated activation energy of photooxidation of different species of titanium oxide is associated with the transport of photoelectron through the catalyst to the adsorbed oxygen on the surface (Harvey et al., 1983 a & b). Kim & Lee, (2010) explained that the very small activation energy in photocatalytic reactions is the apparent activation energy E_a, whereas the true activation energy E_t is nil. These types of reactions are operating at room temperature.

Palmer et al., (2002) observed that the effect of temperature on the photocatalytic degradation is insignificant in the range of 10-68 º C. High temperatures may have a negative impact on the concentration of dissolved oxygen in the solution and consequently, the recombination of holes and electrons increases at the surface of photocatalyst. However, Trillas et al., (1995); Chen & Ray, (1998) reported that raising the temperature of reaction enhances the rate of photocatalytic degradation significantly. Hussein and Abbas (2010 b) reported that the decolorization efficiency of real textile industrial wastewater increases with increasing of temperature as shown in fig. 13.

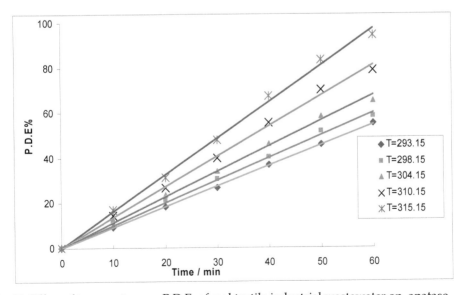

Fig. 13. Effect of temperature on P.D.E. of real textile industrial wastewater on anatase under solar radiation

Hussein et al., (2011) have found that the rate of decolorization of Bismarck brown R on ZnO and different types of TiO$_2$ increases slightly with the increase of the temperature and the activation energy 24 ± 1 kJ.mol^{-1} for ZnO and 14 ± 1, 16 ± 1, 21 ± 1 and 22 ± 1 kJ.mol^{-1} for TiO$_2$ (Degussa P25), TiO$_2$ (Hombikat UV100), TiO$_2$ (Millennium PC105), and TiO$_2$ (Koronose 2073), respectively. Figure 13   shows the impact of temperature on photodecolorization of Bismarck brown R by using TiO$_2$ (Hombikat UV100).

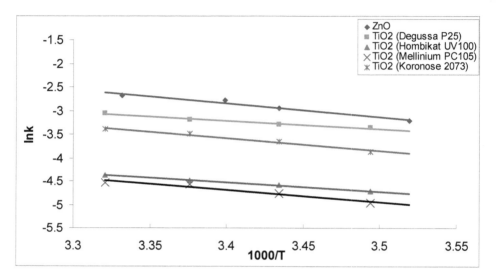

Fig. 14. Arrhenius plot by different types of catalyst with Bismarck brown R

6.6 Effect of addition of oxidants

It is well known that the addition of oxidants increases the rate of photocatalytic degradation of dyes by the formation of hydroxyl radicals (Salvador & Decker, 1984; Jenny & Pichat, 1991). However, this is not general for all types of dyes (Hachem et al., 2001).

Production of additional hydroxyl radicals occurs when hydrogen peroxide is added through the following mechanisms (Galvez, 2003 & Dong et al., 2010)   :

1.   Trapping of photogenerated electrons.

$$H_2O_2 + 2e^- \rightarrow 2OH^-$$
(60)

2.   Self-decomposition by photolysis

$$H_2O_2 + h\nu \rightarrow 2\overset{\bullet}{O}H$$
(61)

3.   Reaction with superoxide radical anion O2$^{\bullet-}$

$$H_2O_2 + O_2^- \rightarrow \overset{.}{O}H + OH^- + O_2 \tag{62}$$

The addition of persulphate leads to form sulphate radical anion by trapping the photogenerated electrons (Konstantinou & Albanis, 2004)

$$S_2O_8^{2-} + e^- \rightarrow SO_4^{2-} + SO_4^- \tag{63}$$

The formed sulphate radical anion is a strong oxidant and reacts with organic molecules pollutants as follows (Galvez, 2003):
1. Abstracting a hydrogen atom from saturated carbon.
2. Adding hydrogen to unsaturated or aromatic carbon.
3. Removing one electron from carboxylate anions and from certain neutral molecules.
Sulphate radical anion can also react with water molecule to produce hydroxyl radical (Konstantinou & Albanis, 2004):

$$SO_4^- + H_2O \rightarrow SO_4^{2-} + \overset{.}{O}H + H^+ \tag{64}$$

Other oxidants such as iodate and bromate can also increase the reaction rate because they are also electron scavengers, while chlorate has been proven insufficient to improve effectiveness (Galvez, 2003). However, these additives are too expensive to be compared to hydrogen peroxide and peroxydisulphate. Moreover, they do not dissociate into harmless products.
The addition of oxidant to reaction mixture serves the rate of photocatalytic degradation by:
1. Generation of additional •OH and other oxidizing species.
2. Increasing the number of trapped photoelectrons.
3. Increasing the oxidation rate of intermediate compounds.
4. Replacement of oxygen role in the case of the absence of oxygen in the reaction mixture.
Table 2 shows the effect of addition of hydrogen peroxide on the rate of photocatalytic degradation of red disperse dye on ZnO. The results indicate that the apparent rate constant increases with the increase in H_2O_2 concentration to a certain level and a further increase in H_2O_2 concentration leads to decrease in the degradation rate of the red disperse dye. The presumed reason is that the addition of H_2O_2 to a certain level increases the production of hydroxyl radicals , but the additional amount leads to reduce the amounts of photoholes and hydroxyl radicals (Legrini et al., 1993; Malato, 1998; Daneshvar et al., 2003; Konstantinou & Albanis, 2004):

$$H_2O_2 + 2h^+ \rightarrow O_2 + 2H^+ \tag{65}$$

$$H_2O_2 + \overset{.}{O}H \rightarrow H_2O + HO_2 \tag{66}$$

$$HO_2 + \overset{.}{O}H \rightarrow H_2O + O_2 \tag{67}$$

Conc. of H_2O_2 (mmol.L^{-1})	K_{app} (min^{-1})
0	0.0893
1.5	0.1408
3.0	0.1499
4.5	0.1680
6.0	0.1404
7.5	0.1290

Table 2. Effect of addition of H_2O_2 on the apparent rate constant of photocatalytic decolorization of red disperse dye.

This behavior relates to the competition between the adsorption of organic pollutants and H_2O_2 on the surface of photocatalyst. The required amount of H_2O_2 reaches the highest level of enhancement for the rate of photodegradation is related to the ratio of the concentration of organic pollutants and H_2O_2 (Galvez, 2003). When the pollutant concentration is low compared with the concentration of H_2O_2, the adsorption of organic pollutants decreases due to the increase of adsorption of hydrogen peroxide and, as a result, the additional hydroxyl radicals generated by H_2O_2 do not react efficiently.

6.7 Comparison between mineralization and photocatalytic decolorization
Mineralization of dyes is a process in which dyes are converted completely into its inorganic chemical components (minerals), such as carbon dioxide, water and other species according to the structure of dye (see figs. 1&2).

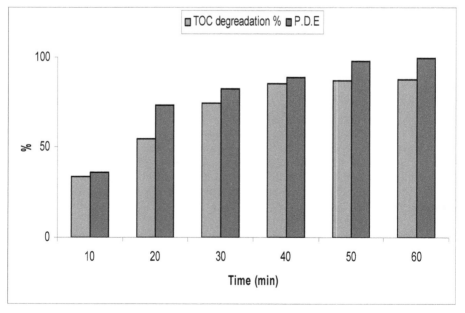

Fig. 15. Comparison between Mineralization and Photocatalytic decolorization of Bismarck brown R on ZnO

Mineralization of Bismarck brown R was evaluated by analyzing total organic carbon (TOC) (Hussein et al. 2011). The results shown in Fig.15 indicate that photocatalytic decolorization of Bismarck R brown was faster than the decrease of TOC. The results indicate that % TOC reduction was about 73% after 60 minutes of irradiation while the per cent of decolorization achieved 88% for the same period of irradiation. These findings in good agreement with those reported before (He et al. 2007 & Chen 2009). This could be explained to the formation of some by product, which resist the photocatalytic degradation. Furthermore, the formed by products need more time to destruct.

6.8 Effect of irradiation sources

Table 3 summarized the obtained results from the different techniques used for the treatments of textile industrial wastewater and different types of industrial dyes (Hussein, 2010). The results show that the decolorization rate of textile industrial wastewater is faster with solar light than with UV light. The results indicated that solar energy could be effectively used for photocatalytic degradation of pollutants in wastewater.

Process	Type of treated waste or dye	Source of irradiation	Type of catalyst	Time for complete mineralization/hours	Reference
Photocatalytic	Textile industrial wastewater	Solar	ZnO	2.7	Alkhateeb et al., (2005)
Photocatalytic	Textile industrial wastewater	Solar	TiO_2	4	Alkhateeb et al., (2005)
Photolysis	Murexide	Solar	-	7.5	Alkhateeb et al., (2007)
Photocatalytic	Murexide	Solar	ZnO	3.8	Alkhateeb et al., (2007)
Photocatalytic	Murexide	Solar	TiO_2	3.3	Alkhateeb et al., (2007)
Photolysis	Thymol blue	Solar	-	7	Hussein et al., (2008)
Photocatalytic	Thymol blue	Solar	TiO_2	2.7	Hussein et al., (2008)
Photocatalytic	Thymol blue	Solar	ZnO	3.3	Hussein et al., (2008)
Photocatalytic	Textile industrial wastewater	Mercury lamp	TiO_2	2.6	Attia et al., (2008)
Photocatalytic	Textile industrial wastewater	Mercury lamp	ZnO	3	Attia et al., (2008)
Photocatalytic	Bismarck brown G	Mercury lamp	ZnO	1	Hussein et al., (2010a)
Photocatalytic	Bismarck brown R	Mercury lamp	ZnO	0.8	Hussein et al., (2010b)
Photocatalytic	Bismarck brown R	Mercury lamp	TiO_2	1.2	Hussein et al., (2010c)
Photocatalytic	Textile industrial wastewater	Mercury lamp	TiO_2	3	Hussein & Abass, (2010a)
Photocatalytic	Textile industrial wastewater	Mercury lamp	ZnO	1	Hussein & Abass, (2010a)
Photocatalytic	Textile industrial wastewater	Solar	TiO_2	1.8	Hussein & Abass, (2010b)
Photocatalytic	Textile industrial wastewater	Solar	ZnO	0.33	Hussein & Abass, (2010b)

Table 3. Effect of irradiation sources on photocatalytic decolorization of textile industrial wastewater and different dyes

7. Conclusions

1. Photocatalytic degradation techniques is the most efficient and clean technology.
2. Textile industries have become worldwide. Thus, this method can be considered as a promising technique for providing formidable quantities of water especially for countries facing serious suffering from water shortage.
3. The existence of catalyst and lights are essential for photocatalytic degradation of colored dyes.
4. Photocatalytic degradation efficiency (PDE) of textile industrial wastewater is obviously affected by illumination time, pH, initial dye concentration and photocatalyst loading.
5. Solar photocatalytic treatment has been proved to be an efficient technique for decolorization of industrial wastewater through a photocatalytic process and the transformation is practically complete in a reasonable irradiation time.
6. In the countries where, intense sunlight is available throughout the year, solar energy could be effectively used for photocatalytic degradation of pollutants in industrial wastewater.
7. The zero point charge is 6.4 and 9.0 for TiO_2 and ZnO, respectively above which the surface of photocatalyst is negatively charged by means of adsorbed hydroxyl ions; this favors the formation of hydroxyl radical, and as a result, the photocatalytic degradation of industrial wastewater increases due to inhibition of the photoholes and photoelectrons recombination.

8. References

Ahmed, S. ; Rasul, M.; Martens W. ; Brown , R.& Hashib, M.(2011). Advances in Heterogeneous Photocatalytic Degradation of Phenols and Dyes in Wastewater: A Review. *Water Air Soil Pollut* , Vol. 215,pp.3–29.

Akyol, A.; Yatmaz, H. & Bayramoglu, M. (2004). Photocatalytic Decolorization of Remazol Red RR in Aqueous ZnO Suspensions. *Applied Catalysis B: Environmental*, Vol. 54, pp. 19–24.

Alkhateeb, A.; Hussein, F. & Asker, K. (2005). Photocatalytic Decolorization of Industrial Wastewater Under Natural Weathering Conditions. *Asian J. Chem.*, Vol. 17, No. 2, pp. 1155-1159.

Al-zahra, F. ; Alkhateeb, A. & Hussein F. (2007). Photocatalytic Oxidation of Benzyl Alcohol Using Pure and Sensitized Anatase, *Desalination*, Vol. 209, pp. 342-349.

Alkhateeb, A. ; Ismail, J. & Hussein, F.(2007). Solar Photolysis and Photocatalytic Degradation of Murexide Using Titanium Dioxide and Zinc Oxide. *J. of Arab University for Basic and Applied Sciences*, Vol. 4, pp. 70-76.

Antharjanam, S.; Philip, R. & Suresh, D.(2003). Photocatalytic Degradation for Wastewater Pollutants, *Ann. Chim.*, Vol. 93, No. 9-10 , pp.719-728.

Attia, A.; Kadhim, S. &. Hussein, F. (2008).Photocatalytic Degradation of Textile Dyeing Wastewater Using Titanium Dioxide and Zinc Oxide, *E-J. Chem.*,Vol. 5,pp. 219-223

Bahnemann, D.; Bockelmann, D. & Goslich, R. (1991). Mechanistic Studies of Water Detoxification on Illuminated TiO_2 Suspension, *Solar Energy Mater.*, Vol. 24, pp.564-583.

Bahnemann, D.; Cunningham, J.; Fox, M.; Pelizzetti, E.; Pichat, P. & Serpone, N. (1994). Photocatalytic Treatment of Water, in: *Aquatic and surface photochemistry*, G. Heiz, R. Zepp and D. Crosby. eds, Lewis Publishers, U.S.A. pp. 261–316.

Bhutani., S.(2008). *Organic chemistry-selected topics*, 1st ed., Ane Book India, 169- 172, New Delhi.

Buxton, G.; Greenstock, C.; Helman, W.& Ross, A. (1988). Critical Review of Rate Constants for Reactions of Hydrated Electrons, Hydrogen Atoms and Hydroxyl Radicals in Aqueous Solution, *J. Phys. Chem. Ref. Data.*, Vol. 17, No.2,pp. 513- 886.

Chen , Chih-Yu , (2009). Photocatalytic Degradation of Azo Dye Reactive Orange 16 by TiO2, *Water Air Soil Pollut* Vol.202,pp. 335–342.

Chen, D. & Ray, A. (1998). Photodegradation Kinetics of 4-nitrophenol in TiO$_2$ Suspension, *Wat. Res.*, Vol. 32, No. 11, pp.3223-3234.

Chen, J.; Liu, M.; Zhang, J. ; Ying, X. & Jin, L. (2004). Photocatalytic Degradation of Organic Wastes by Electrochemically Assisted TiO$_2$ Photocatalytic System, *J. Environ. Manage.*, Vol. 70, No. 1, pp. 43-47.

Cho, Y.; Choi, W.; Lee,C.; Hyeon, T. & Lee, H.(2001).Visible Light-Induced Degradation of Carbon Tetrachloride on Dye-Sensitized TiO$_2$.*Environ. Sci. Technol.*, Vol. 35, pp. 966-970.

Cooper, P. (1995). Removing Colour From Dye House Wastewater. *Asian Textile Journal*, Vol. 3, pp. 52-56.

Correia, V.; Stephenson, T. & Judd, S. (1994). Characterization of Textile Wastewaters, A Review. *Environ. Technol.*, Vol. 15, pp.917–929.

Couto, S. & Toca-Herrera, J.(2006). Lacasses in the Textile Industry, *Mini Review, Biotechnology and Molecular Biology Review.* Vol. 1, No. 4, pp. 115-120.

Daneshvar, N.; Salari, D. & Khataee, A. (2003). Photocatalytic Degradation of Azo Dye Acid Red 14 in Water: Investigation of The Effect of Operational Parameters. *Journal of Photochemistry and Photobiology* A, Vol. 157, pp. 111–116.

Domenech, X.; Jardim, W. F. & Litter, M. I. (2001). Procesos Avanzados De Oxidacion Para La Eliminacion De Contaminantes. Cap 01 Do Livro Eliminacion De Contaminantes Por Fotocatalisis Heterogenea, Editado Por Miguel A. Blesa (Para Cyted), (In Spanish).

Dong, D. ; Peijun, Li ; Li, X.; Zhao, Q.; Zhang, Y.;Jia, C. & Li , P. (2010). Investigation on The Photocatalytic Degradation of Pyrene on Soil Surfaces Using Nanometer Anatase TiO$_2$ Under UV Irradiation. *J. Hazard. Mater.*, Vol. 174, pp. 859–863.

Egerton, T. & King, C. (1979). The Influence of Light Intensity on Photoactivity in TiO$_2$ Pigmented System, *J. Oil. Col. Chem. Assoc.*, Vol. 62, pp.386-391.

Fernandez-Ibanez, P.; Planko, J.; Maitato, S. & de las Nieres, F.(2003). Application of Colloidal Stability of TiO$_2$ Particles for Recovery and Reuse in Solar Photocatalysis, *Water Res.*, Vol. 37, No. 13 , pp.3180-3188.

Galvez, J.(2003). *Solar Detoxification*, United Nations Educational, Scientific and Cultural Organization, Electronic Copy.

Gaya, U.; Abdullah, A. ; Zainal, Z .& Hussein, M. (2010). Photocatalytic Degradation of 2,4-dichlorophenol in Irradiated Aqueous ZnO Suspension, *International Journal of Chemistry*, Vol. 2, No. 1, pp.180-193.

Glaze, W.; Kang. J. & Chapin, D. (1987). The Chemistry of Water Treatment Processes Involving Ozone, Hydrogen Peroxide and Ultraviolet Radiation. *Ozone Sci. and Eng.*, Vol. 9, pp.335-342.

Gomes da Silva & Cfaria, J. (2003). Photochemical and Photocatalytic Degradation of an Azo Dye in Aqueous Solution by UV Irradiation. *J. Photochem. Photobiol. A: Chem.*, Vol. 155, pp. 133-143.

Guillard, C. ; Puzenat, E.; Lachheb, H.; . Houas, A. & Herrmann, J. (2005). Why Inorganic Salts Decrease the TiO_2 Photocatalytic Efficiency, *International Journal of Photoenergy*, Vol. 7, pp.1-9.

Hachem, C.; Bocquillon, F.; Zahraa, O. & Bouchy, M.(2001). Decolourization of Textile Industry Wastewater by the Photocatalytic Degradation Process. *Dyes and Pigments*, Vol.49, pp. 117–125.

Haque, M. & Muneer, M. (2007). TiO_2-mediated Photocatalytic Degradation of a Textile Dye Derivative, Bromothymol Blue, in Aqueous Suspensions, *Dyes and Pigments*, Vol. 75, pp. 443-448.

Harvey, P.; Rudham, R. & Ward, S.(1983 a). Photocatalytic Oxidation of Liquid Propan-2-0l by Titanium Dioxide, *J. Chem. Soc. Faraday Trans.* 1 . Vol. 79, pp. 1381-1390.

Harvey, P.; Rudham, R. & Ward, S. (1983 b). Photocatalytic Oxidation of Liquid Alcohols and Binary Alcohol Mixtures by Rutile, *J. Chem. Soc. Faraday Trans.* 1. Vol. 79, pp.2975-2981.

He, Z. ; Song, S. ; Zhou, H. ; Ying, H. & Chen, J. (2007). C.I. Reactive black 5 decolorization by combined sonolysis and ozonation. *Ultrasonics Sonochemistry*, Vol. 14,pp. 298–304.

Hird, M. (1976). Transmission of Ultraviolet Light by Films Containing Titanium Pigments - Applications in UV Curing. *J. Coatings Tech.*, Vol.48, pp 75-82.

Hussein, F.(1984). Photocatalytic Dehydrogenation of Liquid Alcohols by Platinized Anatase and other Catalysts, PhD Thesis, Nottingham University, UK.

Hussein, F. (2002). Photocatalytic Oxidation of Liquid Alcohols by Titanium Dioxide, Abhath Al-Yarmouk, *Basic Sciences and Engineering*, Vol. 11, No. 1B, pp.327-336.

Hussein, F. (2010). Water Availability in IRAQ and Recycling of Wastewater, The 11th International Forum on Marine Science & Technology and Economic Development, Asian-Pacific Conference on Desalination and Water Reclamation, China, pp. 388-394, Qingdao, China , June 22-25, 2010.

Hussein ,F.; Pattenden, G.; Rudham, R. & Russell, J. (1984). Photo-Oxidation of Alcohols Catalysed by Platinised Titanium Dioxide, *Tetrahedron Letters,* Vol.25, No.31, pp 3363-3364.

Hussein, F. ; Radi S. & Naman, S.(1991). The Effect of Sensitizer on the Photocatalytical Oxidation of Propan-2-ol by Pt-TiO_2 and other Catalysts, Energy and Environmental Progress 1, Vol. B, Solar Energy Applications, Bioconversion and Sunfules, Nova Science Publisher, Inc. USA, ISBN 0-941743-97-7, pp. 337-353.

Hussein, F. & Rudham, R.(1984). Photocatalytic Dehydrogenation of Liquid Propan-2-0l by Platinized Anatase and Other Catalysts, *J. Chem. Soc. Faraday Trans.* 1. Vol. 80, pp.2817-2825.

Hussein, F. & Rudham, R.(1987). Photocatalytic Dehydrogenation of Liquid Alcohols by Platinized Anatase, *J. Chem. Soc., Faraday Trans.* 1. *Vol. 83.* pp.1631-1639.

Hussein, F. &. Alkhateeb, A. (2007), Photo-oxidation of Benzyl Alcohol under Natural Weathering Conditions, *Desalination* ,Vol.209,pp.350-355.

Hussein, F. & Abbas, T.(2010 a). Photocatalytic Treatment of Textile Industrial Wastewater, *Int. J. Chem. Sci.* Vol. 8, No. 3, pp. 1353-1364.

Hussein, F. & Abbas, T.(2010 b). Solar Photolysis and Photocatalytic Treatment of Textile Industrial Wastewater, *Int. J. Chem. Sci.,* Vol. 8, No. 3, pp. 1409-1420.

Hussein, F.; Alkhateeb, A. & Ismail, J.(2008). Solar Photolysis and Photocatalytic Decolorization of Thymol Blue, *E-J. Chem.*, Vol. 5, No. 2, pp.243-250.

Hussein, F.; Halbus, A. ; Hassan, H. & Hussein, W.(2010 a). Photocatalytic Degradation of Bismarck Brown G using Irradiated ZnO in Aqueous Solutions, *E-J. Chem.*, Vol. 7, No. 2, pp.540-544.

Hussein, F. ; Obies, M. ; & Drea, A. (2010 b), Photocatalytic Decolorization of Bismarck Brown R by Suspension Of Titanium Dioxide, *Int. J. Chem. Sci.*, Vol.8,No.4,pp. 2736-2746.

Hussein, F. ; Obies, M. ; & Drea, A. (2010 c), Photodecolorization of Bismarck Brown R in The Presence of Aqueous Zinc Oxide Suspension, *Int. J. Chem. Sci.*, Vol. 8, No. 4 ,pp. 2763-2774.

Hussein, F. ; Obies, M. & Bahnemann, D. (2011), Photocatalytic Degradation of Bismarck Brown R Using Commercial ZnO and TiO_2, to be published later.

Jenny, B. & Pichat, P. (1991). Determination of The Actual Photocatalytic Rate of H_2O_2 Decomposition Over Suspended TiO_2. Fitting to the Langmuir–Hinshelwood form, *Langmuir*, Vol.7, pp. 947–54.

Kavitha, S. & Palanisamy, P. (2011). Photocatalytic and Sonophotocatalytic Degradation of Reactive Red 120 Using Dye Sensitized TiO_2 under Visible Light, *International Journal of Civil and Environmental Engineering*,Vol. 3, No. 1,pp.1-6.

Kim, T. & Lee M. (2010). Effect of pH and Temperature for Photocatalytic Degradation of Organic Compound on Carbon-coated TiO_2, *J. of Advanced Engineering and Technology*, Vol. 3, No. 2, pp. 193-198.

Konstantinou, I. & Albanis,T. (2004). TiO_2-assisted Photocatalytic Degradation of Azo Dyes in Aqueous Solution: Kinetic and Mechanistic Investigations . A review. *Appl. Catal. B: Environ.*, pp. 49 1–14.

Kormann, C.; Bahnemann, D. & Hoffmann M.(1991). Photolysis of Chloroform and Other Organic Molecules in Aqueous TiO_2 Suspensions, *Environ. Sci. Technol.*, Vol. 25, pp. 494-500.

Legrini, O.; Oliveros, E. & Braun, A. (1993). Photochemical Processes for Water Treatment, *Chem. Rev.*, Vol. 93, pp. 671-698.

Li, X.; Liu, H.; Cheng, L. & Tong, H.(2003). Photocatalytic Oxidation Using a New CatalystsTiO_2 Microspheresfor Water and Wastewater Treatment. *Environ. Sc. Technol.*, Vol. 37, No. 37 ,pp. 3989-3994.

Liu HL-Zhou, D.; Li, X. & Yue, P.(2003). Photocatalytic Degradation of Rose Bengal. *J. Environ. Sci.* (China). Vol. 15, No. 5 , pp. 595-597.

Liu, Y.; Chen, X.; Li, J. & Burda, C. (2005). Photocatalytic Degradation of Azo Dyes by Nitrogen-doped TiO_2 Nano Catalysts. Chemosphere, Vol. 61, pp 11–18.

Malato, S.; Blanco, J.; Richter, C.; Braun, B. & Maldonado M. (1998). Enhancement of the Rate of Solar Photocatalytic Mineralization of Organic Pollutants by Inorganic Oxidising Species. *Appl. Catal. B: Environ.*,Vol. 17, pp.347-360.

Mota, A. L. N.; Albuquerque, L. F.; Beltrame, L. T. C.; Chiavone-Filho, O.; Machulek Jr., A. & Nascimento, C. A. O. (2008) "Advanced Oxidation Processes and Their Application in the Petroleum Industry: A Review". *Brazilian Journal Of Petroleum And Gas*. Vol. 2, No. 3, pp. 122-142.

Mutambanengwe, C. (2006). Hydrogenases from Sulphate Reducing Bacteria and Their Role in the Bioremediation of Textile Effluent, MSc Thesis, Rhodes University PP.15.

Obies, M. (2011). Photocatalytic Decolorization of Bismarck Brown R, MSc Thesis , Chemistry Department, College of Science, Babylon University, Iraq.

Ohno, T.(2004). Preparation of Visible Light Active S-doped TiO_2 Photocatalysts and Their Photocatalytic Activities, *Water Sci. Technol.*, Vol.49, No. 4 , pp. 159-163.

Ollis, D. (1991). Solar-Assisted Photocatalysis for Water Purification: Issues, Data, Questions. Photochemical Conversion and Storage of Solar Energy, pp. 593-622, Kluwer Academic Publishers.

Orhon, D.; Kabdasli, I.; Germirli Babuna, F.; Sozen, S.; Dulkadiroglu, H.; Dogruel, S.; Karahan, O. & Insel, G.(2003). Wastewater Reuse for The Minimization of Fresh Water Demand in Coastal Areas-selected Cases from The Textile Finishing Industry. *J. Environ. Sci. Health* . Vol. A 38, pp.1641– 1657.

Palmer, F.; Eggins, B. & Coleman, H., (2002). The Effect of Operational Parameters on The Photocatalytic Degradation of Humic Acid. *J. Photochem. Photobiol. A.Chem.*, Vol.148, No. 1-3, pp.137-143.

Peterson, M. ; Turner, J. & Nozik, A. (1991). Mechanistic Studies of The Photocatalytic Behavior of TiO_2 : Particles in a Photoelectrochemical Slurry cell and relevance to Photo Detoxification Reactions. *J. Phys. Chem.*, Vol.95, pp.221- 225 .

Pignatello, J.; Oliveros, E. & MacKay, A. (2006). Advanced Oxidation Processes for Organic Contaminant Destruction Based on The Fenton Reaction and Related Chemistry. *Crit. Rev. Environ. Sci. Technol.*, Vol. 36, pp. 1-84.

Sakthivela, SB.; Neppolianb ,MV.; Shankarb, B.; Arabindoob, M.; Palanichamyb, V. & Murugesanb, V. (2003). Solar Photocatalytic Degradation of Azo Dye: Comparison of Photocatalytic Efficiency of ZnO and TiO2, *Sol. Energy Mater. Sol. Cells*, Vol. 77, pp. 65–82.

Salvador, P. & Decker, F. (1984). on The Generation of H_2O_2 During Water Photoelectrolysis at n-TiO_2. *J Phys Chem.*,Vol. 88, PP. 6116–20.

Shaw, C.; Carliell, C. & Wheatley, A. (2002). Anaerobic/aerobic Treatment of Coloured Textile Effluents Using Sequencing Batch Reactors. *Water Res.*, Vol. 36, pp. 1993 – 2001.

Stephen, J. (1995). Electrooxidation of Dyestuffs in Waste Waters. *J. Chem. Technol. Biot.*, Vol. 62, pp. 111-117.

Stylidi, M.; Kondarides, D. & Verykios, X. (2003). Pathways of Solar Light-Induced Photocatalytic Degradation of Azo Dyes in Aqueous TiO_2 Suspensions. *Appl. Catal. B Environ.*, Vol. 40, No. 4, pp.271-286.

Szczepantiewicz, S. ; Colussi, A. & Hoffmann, M. (2000). Infrared Spectra of Photoinduced Species on Hydroxylated Titania Surface, *J. Phys. Chem.* B., Vol. 104, pp.9842-9850.

Trillas, M.; Peral, J. & Domènech, X. (1995). Redox Photodegradation of 2,4-dichlorophenoxyacetic Acid Over TiO_2. *Appl. Catal. B Environ.*, Vol. 5, No. 4, pp.377-387.

Vione, D.; Picatonitoo, T. & Carlotti, M.(2003). Photodegradation of Phenol and Salicylic acid by Coated Rutile Based Pigment, *J. Cosmet Sci.*, Vol. 54,pp. 513-524.

Weast, R; (Ed.) (1977). *Handbook of Chemistry and Physics*, 58th edition, CRC Pres.

Zhao M. & Zhang J.(2008), Wastewater Treatment by Photocatalytic Oxidation of Nano-ZnO. *Global Environmental Policy in Japan*, No.12 ,pp.1-9.

4

Effect of Photochemical Advanced Oxidation Processes on the Bioamenability and Acute Toxicity of an Anionic Textile Surfactant and a Textile Dye Precursor

Idil Arslan-Alaton and Tugba Olmez-Hanci
Istanbul Technical University
Turkey

1. Introduction

Surfactants are frequently being used as cleaning, dissolving and wetting agents in household activities and several industries (Utsunomiya et al., 1997; Van de Plassche et al., 1999). For instance in various textile preparation operations (scouring, mercerising, bleaching), surfactant formulations are employed in order to allow thorough wetting of the textile material, emulsification of lipophilic impurities and dispersion of insoluble matter and degradation products (Arslan-Alaton et al., 2007; EU, 2003). Anionic (including alkyl sulphonates, alkyl aryl sulphonates, alkyl sulphates, dialkylsulphosuccinates, and others) and especially nonionic surfactants are the chemicals being more often used for this particular purpose. These multi-purpose surfactants create the main organic pollution load in effluents originating from the above mentioned activities. Surfactants enter the environment mainly through the discharge of sewage effluents into natural water and the application sewage sludge on land for soil fertilizing purposes (Petrovic & Barceló, 2004). Although most of these surfactants are designated as "biodegradable" according to different long-term biodegradability tests (Euratex, 2000) former studies dealing with the biodegradability of surfactants have indicated that their biodegradation in conventional activated sludge treatment systems is in most cases rather incomplete, resulting in the accumulation of partial biodegradation products (Ikehata & El-Din, 2004; Swisher, 1987). A lot of commercial surfactants used today by different industries tend to sorb and hence accumulate on sludge and soil sediments in receiving water bodies due to their amphiphilic characteristics (Swisher, 1987; Staples et al., 2001). Hence, when a surfactant is discharged into the environment without proper treatment at source (e.g. at the treatment facilities of the textile factory), it will enter the sewage treatment works or directly natural waters without any significant structural alteration and may cause serious ecotoxicological consequences due to its bioaccumulation tendency (Ikehata & El-Din, 2004). Consequently, more effective and at the same time economically feasible treatment processes have to be applied to alleviate the problem of partially adsorbed and/or metabolized surfactants in the environment. Moreover, the management of biologically-difficult-to-degrade effluent discharges bearing surfactants remains an important challenge that has to be urgently solved to reduce the concentration of surfactants in effluent discharge at source.

Aryl sulphonates are commercially important textile dye precursors; their global production for dye synthesis is limited to a few countries (e.g. China, India, Pakistan and Egypt) where they create serious environmental problems since they are biologically inert in natural as well as engineered aerobic biotreatment systems. Moreover, due to their very polar and hydrophilic nature, sulphonated textile dye precursors are very mobile in the aquatic ecosystem and hence very difficult to treat via conventional methods (i.e. coagulation, precipitation, adsorption). In addition, it has been postulated that the anaerobic metabolites of sulphonates bearing aromatic amines in their molecular structure are potentially mutagenic and even carcinogenic. From the environmental point of view, the fate of aryl sulphonates and their degradation products in biological treatment units as well as receiving water bodies is still not very clear since until now only limited attention has been paid towards their occurrence, fate and degradability in engineered systems as well as in the natural environment (Jandera et al., 2001). Due to the fact that conventional biological, physical and chemical treatment methods are not very effective in the removal of naphthalene sulphonates, alternative options have to be considered and evaluated.

In the last three decades, so-called Advanced Oxidation Processes (AOPs) have been developed and explored for the treatment of toxic and/or refractory pollutants and wastewaters. Their efficiency and oxidation power relies on the formation of very active species including hydroxyl radicals (HO•) that violently and nonselectively react with most organic pollutants at diffusion limited rates. Bimolecular rate coefficients obtained for the free radical chain reaction initiated by HO• are typically in the range of 10^7-10^{11} M^{-1} s^{-1} for most pollutants depending on their structural properties. Among the AOPs, photochemical processs including the UV-C photolysis of H_2O_2 and the Photo-Fenton reaction (UV-C photolysis of H_2O_2 catalyzed by Fe^{2+} ions under acidic pH conditions) are known as the most efficient, feasible and kinetically favorable types. The major drawbacks of AOPs can be listed as (i) their relatively high electrical energy requirements dramatically affecting their operating costs and hence real-scale application as well as (ii) the possibility of undesired oxidation products (more toxic and less bioamenable than the mother pollutant) formation and build-up due to kinetically difficult to control chain reactions in the treatment system. Baxendale and Wilson (1957) first studied the UV-C photolysis of H_2O_2 in water that is the most direct method of HO• generation through the homolytic cleavage of H_2O_2. The following set of reactions explains the main steps of the mechanism of the H_2O_2 photolytic decomposition in water;

$$H_2O_2 + hv \rightarrow 2HO^\bullet \tag{1}$$

$$HO^\bullet + H_2O_2 \rightarrow HO_2^\bullet + H_2O \tag{2}$$

$$HO^\bullet + HO_2^- \rightarrow HO_2^\bullet + OH^- \tag{3}$$

$$2HO^\bullet \rightarrow H_2O + \tfrac{1}{2}O_2 \tag{4}$$

The quantum yield of reaction (4) is 1.0 mole HO• per mole of photons (1.0 einstein) absorbed by H_2O_2 (Prousek, 1996). The molar absorption coefficient of H_2O_2 at 254 nm is only 18.6 M^{-1} cm^{-1} and hence the efficiency of the H_2O_2/UV-C process decreases drastically with increasing pollutant concentration and UV_{254} absorbance of the target chemical (CCOT 1995). For heavily polluted effluents, high UV-C doses and/or H_2O_2 concentrations are required.

The Fenton process is being increasingly used in the treatment of organic pollutants. The conventional dark Fenton process involves the use of one or more oxidizing agents (usually H_2O_2 and/or oxygen) and a catalyst (a metal salt or oxide, usually iron) at acidic pHs. Fenton process is efficient only in the pH range 2-4 and is usually most efficient at around pH 2.8 (Pignatello, 1992). In acidic water medium, the most accepted scheme of this radical mechanism is described in the following equations (Sychev & Isak, 1995):

$$Fe^{2+}+H_2O_2 \rightarrow Fe^{3+}+HO^-+HO^\bullet \tag{5}$$

$$Fe^{3+}+H_2O_2 \rightarrow Fe^{2+}+HO_2^\bullet+H^+ \tag{6}$$

$$Fe^{2+}+HO^\bullet \rightarrow Fe^{3+}+HO^- \tag{7}$$

$$HO^\bullet+H_2O_2 \rightarrow HO_2^\bullet+H_2O \tag{8}$$

$$HO_2^\bullet \leftrightarrow O_2^{\bullet-}+H^+ \quad K=1.58 \times 10^{-5} \text{ M} \tag{9}$$

$$Fe^{3+}+HO_2^\bullet/O_2^{\bullet-} \rightarrow Fe^{2+}+O_2/+H^+ \tag{10}$$

$$Fe^{2+}+HO_2^\bullet/O_2^{\bullet-} \rightarrow Fe^{3+}+H_2O_2 \tag{11}$$

Photo-Fenton oxidation is the photo-catalytically enhanced version of the Fenton process; in the Photo-Fenton process; UV light irradiation (180-400 nm) causes an increase in the $HO^\bullet$ formation rate and efficiency via photoreduction of Fe^{3+} to Fe^{2+}, thus continuing the redox cycle as long as H_2O_2 is available (Pignatello, 1992; Wadley & Waite, 2004). The reason for the positive effect of irradiation on the degradation rate include the photoreduction of Fe^{3+} to Fe^{2+} ions, which produce new $HO^\bullet$ with H_2O_2 (Eq. 6) according to the following mechanism;

$$Fe^{3+}+H_2O \xrightarrow{h\upsilon} Fe^{2+}+HO^\bullet+H^+ \tag{12}$$

$$H_2O_2 \xrightarrow{h\upsilon} 2HO^\bullet \quad (\lambda<400 \text{ nm}) \tag{13}$$

The main compounds absorbing UV light in the Fenton system are ferric ion complexes, e.g. $[Fe^{3+}(OH)^-]^{2+}$ and $[Fe^{3+}(RCO_2)^-]^{2+}$, which produce additional Fe^{2+} by the following photo-induced, ligand-to-metal charge-transfer reactions (Sagawe et al., 2001);

$$\left[Fe^{3+}(OH)^-\right]^{2+} \rightarrow Fe^{2+}+HO^\bullet \quad (< \text{ca. 450 nm}) \tag{14}$$

$$\left[Fe^{3+}(RCO)^-\right]^{2+} \rightarrow Fe^{2+}+CO_2+R^\bullet \quad (< \text{ca. 500 nm}) \tag{15}$$

Additionally, reaction (14) yields $HO^\bullet$, while reaction (15) results in a decrease of the total organic carbon (TOC) content of the system due to the decarboxylation of organic acid intermediates. It is very important to note that both reactions produce ferrous ions required for the Fenton reaction (Eq. 6). The overall degradation rate of organic compounds is considerably increased in the Photo-Fenton process, even at lower concentration of iron salts present in the system (Chen & Pignatello, 1997). The main disadvantage of the iron based

AOPs is the necessity to work at low pH, because at higher pH ferric ions would begin to precipitate as hydroxide. Furthermore, depending on the iron concentration used, it has to be removed after the treatment in agreement with the regulation established for wastewater discharge (Rodríguez, 2003).

Information about the toxicity of organic pollutants and advanced oxidation intermediates appears to be a key factor in order to evaluate the applicability of the AOPs. It should be considered that advanced oxidation may result in the formation of more inert and/or toxic degradation products, since free radical chain reaction-based treatment methods are difficult to control thus requiring the toxicity assessment of the formed advanced oxidation intermediates (Farré et al., 2007; Oller et al., 2007; González et al., 2007). Furthermore, in most cases the final goal of AOPs is to improve the biocompatibility of the effluent in order to apply a biological treatment (Scott & Ollis, 1995; Marco et al., 1997). Consequently, the use of short-term biological screening tests that aim to a rapid estimation of the acute toxicity of a pollutant and its oxidation products during AOP applications has become essential. Until now, several studies investigating relative changes in the toxicity of model pollutants have been reported using photobacteria, cladocerans, microalgae and other test organisms (Wang et al., 2002; Dalzell et al., 2002; Stasinakis et al., 2008). Among the available acute toxicity test protocols, the assessment of respirometric inhibition (decrease in oxygen uptake rate of microorganisms relative to a control sample; e.g. synthetic or real sewage) is an attractive and advantageous option since heterotrophic biomass being present in biological treatment systems are used as the toxicity test organism that enables the simultaneous assessment of acute toxicity and biodegradability by a single test protocol (Paixão et al., 2002; Gendig et al., 2003; Henriques et al., 2005; Stasinakis et al., 2008; Olmez-Hanci et al., 2009).

Bearing the above-mentioned facts in mind, the present work aimed at investigating the photochemical advanced oxidation of a commercially important anionic textile surfactant (dodecyl sulpho succinate; DOS) and a frequently used vinyl sulphone dye intermediate, namely Para base (aniline-4-beta-ethyl sulphonyl sulphate-2-sulphonic acid; PB) employing H_2O_2/UV-C and Photo-Fenton treatment processes. Treatment performances of the H_2O_2/UV-C and Photo-Fenton processes were evaluated in terms of DOS, PB, COD and TOC removals. Treatment efficiencies, pseudo-first-order reaction rate coefficients and associated electrical energy requirements were comparatively evaluated for DOS and PB degradation with the H_2O_2/UV-C and Photo-Fenton processes to decide for the technically and economically more attractive treatment solution for each chemical. In the second part of the study, activated sludge inhibition experiments were conducted with untreated and photochemically pretreated DOS and PB samples in order to examine the inhibitory effect of the model pollutant and advanced oxidation intermediates on the respirometric activity of activated sludge microorganisms.

2. Materials and methods

2.1 Chemicals and reagents

DOS and PB model compounds were supplied by a local textile chemical company and used as received without further purification. Some physicochemical and environmental characteristics of the studied model compounds were summarized in Table 1.

Property / Pollutant	DOS	PB
Molecular Formula	$C_{20}H_{37}NaO_7S$	$C_8H_{11}O_9S_3$
Molecular Weight (g mol^{-1})	444	347
g COD(g chemical)$^{-1}$	1.54	1.04
g TOC(g chemical)$^{-1}$	0.69	0.28
Molecular Structure		

Table 1. Physicochemical properties of the anionic surfactant (DOS) and the vinyl sulphone dye precursor (PB). The COD and TOC equivalencies shown in Table 1 were experimentally obtained by preparing individual calibration curves for varying DOS and PB concentrations in water and determination of the corresponding COD and TOC values.

35% w/w H_2O_2 (Fluka, USA) was used as received without any dilution. Residual H_2O_2 was destroyed with enzyme catalase derived from *Micrococcus lysodeikticus* (100181 U mL^{-1}, Fluka, USA). The ferrous iron catalyst source was prepared for daily use by dissolving $FeSO_4.7H_2O$ (Fluka, USA) in distilled water to obtain a 10% (w/v) stock solution. Several concentrations of HNO_3 (Merck, Germany) and NaOH (Merck, Germany) solutions were used for pH adjustment. HPLC-grade acetonitrile (Merck, Germany) was used as the mobile phase in the HPLC measurements. All other reagents were analytical grade and used without purification.

2.2 UV-C photoreactor and light source

All photochemical treatment experiments were performed using a 3250 mL-capacity batch stainless steel batch reactor (length=84.5 cm; width=8 cm) equipped with a 40W low-pressure, mercury vapor sterilization lamp that was located at the center of the reactor in a quartz glass envelope. The incident light flux of the UV lamp and effective light path length were determined via H_2O_2 actinometry (Nicole et al., 1990) as 1.6 x 10^{-5} einstein L^{-1} s^{-1} and 5.1 cm, respectively, at 254 nm. During the photochemical experiments, the reaction solution was continuously circulated through the UV-C photoreactor using a peristaltic pump at a rate of 400 mL min^{-1}. For the H_2O_2/UV-C experiments, H_2O_2 was added to the pH-adjusted aqueous DOS and PB solutions and the reaction mixture was fed to the photoreactor via a peristaltic pump. After the sample t=0 was taken, the reaction was initiated by turning on the UV-C lamp. In case of Photo-Fenton oxidation experiments, after adjusting the initial pH of the reaction solutions to 3.0 ± 0.1, e.g. the optimum pH for Photo-Fenton oxidation (Oliveros et al., 1997), Fe^{2+} catalyst was added. In order to eliminate the effect of the dark Fenton reaction, the other reactant (H_2O_2) was added to the reaction mixture at the very end of the suction period. As the whole solution was fed into photoreactor, UV-C lamp was turned on and photo-oxidation was started. For photochemical oxidation experiments, samples were taken at regular time intervals for analyses up to 180 min.

2.3 Experimental approach

For all photochemical oxidation experiments aqueous solutions of DOS and PB were individually prepared and adjusted to an initial COD of 450 mg L^{-1} since a typical textile preparation and textile dye synthesis wastewater exerts a COD in the range of 400-500 mg L^{-1} (Zollinger, 2001). It is known that an optimum H_2O_2 concentration exists for the H_2O_2/UV-C treatment system depending upon the reaction pH, initial organic carbon content, type and molecular structure of the pollutant under investigation (Muruganandham & Swaminathan, 2004; Arslan-Alaton & Erdinc, 2006). In our preliminary experiments the H_2O_2 concentrations (30 mM) were initially selected close to the stoichiometric oxygen equivalent of the aqueous solutions of DOS and PB (=COD × 2.12 g H_2O_2/g O_2 ≈ 28 mM H_2O_2). For H_2O_2/UV-C treatment of DOS at 30 mM H_2O_2 concentration resulted in significant degradation both in DOS and organic carbon content. However, the theoretically established H_2O_2 concentration of 30 mM did not result in significant organic carbon removals for PB (data not shown). Considering our preliminary results, it was decided to increase the initial H_2O_2 concentration to 60 mM for H_2O_2/UV-C treatment of PB. The initial reaction pH was selected as 10.5 for H_2O_2/UV-C treatment of aqueous DOS solutions, since the effluent pH of a typical textile preparation is in the range 10–11 (EU, 2003), and in previous studies it was observed that the H_2O_2/UV-C oxidation process was practically pH-independent, and no significant changes in the reaction rate constants were found over a wide pH range of 3-11 (Arslan-Alaton & Erdinc, 2006; Arslan-Alaton et al., 2007). H_2O_2/UV-C treatment of aqueous PB solutions was investigated at a reaction pH of 3.0.

For Photo-Fenton experiments, H_2O_2 (30 mM) and Fe^{2+} (0.5 mM) concentrations were selected upon consideration of optimum Fe^{2+}: H_2O_2 molar ratios obtained in former related research work and our own experiences in similar case studies (Arslan-Alaton et al., 2009a; 2009b; 2010). Photo-Fenton experiments were conducted at a fixed pH value of 3.0 ± 0.1 that is generally accepted as the optimum pH for iron-based AOPs (Pignatello, 1992).

As being expected for photochemical AOPs (Oppenländer, 2003), the degradation of the original pollutants (in our case DOS and PB) is usually faster than that of their oxidation products, which are typically expressed in terms of the collective parameters COD and TOC. It was also taken into account that treatment conditions may vary for the removal of original pollutant and its degradation products that are usually collectively presented with the environmental sum parameters COD and TOC. Thus in the present study, the process efficiencies were assessed both in terms of DOS and PB, COD and TOC abatement rates. Residual (unreacted) H_2O_2 and pH were also recorded during the entire treatment period. The pH and ionic strength of the synthetic solutions were not controlled throughout the experiments. All experiments were conducted at a constant temperature of 20°C.

As aforementioned, information about the toxicity of degradation intermediates as well as original pollutants appears to be a key issue for the application of AOPs in water and wastewater treatment as well as integrated chemical and biochemical treatment options (Scott & Ollis, 1995; Farré et al., 2007). By considering this important issue in the last stage of the experimental work, relative changes in the inhibitory effect of DOS and PB photochemical degradation intermediates on heterotrophic biomass (sewage sludge) during the application of photochemical oxidation were also examined.

2.4 Activated sludge inhibition experiments

The procedure of the activated sludge inhibition test was based on the test protocol given in the ISO 8192 Method (2007). This method describes the way to assess the inhibitory effects of a test substance on oxygen consumption of activated sludge (resembling sewage) by measuring the respiration rate under defined conditions in the presence of a known biodegradable substrate and different concentrations of the test substance (Ubay Cokgor et al., 2007). The activated sludge used as the heterotrophic biomass was obtained from a municipal tertiary wastewater treatment located in Istanbul and daily fed with synthetic

Effect of Photochemical Advanced Oxidation Processes on the Bioamenability and Acute
Toxicity of an Anionic Textile Surfactant and a Textile Dye Precursor

87

sewage in accordance with the same test protocol (ISO 8192, 2007).The heterotrophic biomass concentration in the activated sludge was determined according to Standard Methods (APHA/AWWA/WEF, 2005) and expressed as mg L^{-1} of mixed liquid volatile suspended solids (MLVSS). All experiments were run at constant heterotrophic active biomass concentration of 1500 mg L^{-1} to obtain appropriate oxygen uptake rates (OUR) of around 100 mg L^{-1} h^{-1}.The COD of the synthetic sewage was adjusted to 480 mg L^{-1} and used as the "readily biodegradable substrate" in the activated sludge inhibition experiments. Any unreacted H_2O_2 remaining in the reaction samples was destroyed with catalase enzyme and the pH of the solutions was adjusted to 7.0±0.2 before conducting the toxicity tests in order to eliminate potential inhibitory effects of hydrogen peroxide and pH, respectively. The blank control was prepared using the same amount of activated sludge and synthetic sewage solution as in the test dilutions, but without adding untreated or treated aqueous DOS and PB samples. The decrease in dissolved oxygen concentration in the blank control and reaction samples was periodically monitored using a WTW InolabOxi Level 2 oxygen meter at different incubation times (15-180 min). OUR values were calculated on the basis of the linear part of the decreasing dissolved oxygen concentration curves versus test time. Percent inhibition of oxygen consumption (I_{OUR}), for every test sample was calculated using the following equation;

$$I_{OUR}(\%) = \left[(R_B - R_T) \times 100 \right] / R_B \qquad (16)$$

where R_B stands for the OUR in the sample blank and R_T is the OUR in the sample effluent mixture, respectively The I_{OUR} values were, thereafter, plotted against the logarithm of the DOS and PB concentrations. The DOS and PB concentrations resulting in percent decrease in OUR's (in mg L^{-1} DOS or PB) after 15 min incubation period was eventually calculated by interpolation of the "log DOS" or "log PB" versus percent "I_{OUR}" plots obtained for different DOS and PB concentrations.

2.5 Analytical procedures

Reaction samples were analyzed for DOS-PB, COD, TOC, residual H_2O_2 and pH before, during and after photochemical treatment. CODs were determined by the closed reflux titrimetric method according to ISO 6060 (1989). In order to prevent the positive interference of H_2O_2 to COD analyses, the pH of each sample solution was adjusted to 6.5-7.5 and thereafter catalase enzyme made from *Micrococcus lysodeikticus* (1AU destroys 1 μmol H_2O_2 at pH=7, 100181 U mL^{-1}, Fluka grade) was added to destroy any residual H_2O_2.

TOC was monitored on a Shimadzu VPCN model carbon analyzer (combustion method) equipped with an autosampler. Residual H_2O_2 in the samples was determined titrimetically by employing the molibdate-catalyzed iodometric method (Official Methods of Analysis, 1980).

The amount of DOS and PB in the aqueous solutions was measured by high-performance liquid chromatography (HPLC, Agilent 1100 Series, USA). DOS was determined with a fluorescence detector (FLD; λ_{ex} = 225 nm, λ_{em} = 295 nm) and a C8 column. 3 mM NaCl / CH_3CN (80/20; v/v) served as the mobile phase (flow rate of 1.5 mL min^{-1}) for DOS. The instrument detection limit for DOS (1.5 mg L^{-1}) was determined as the lowest injected standard that gave a signal-to-noise ratio of at least 3 and an accuracy of 80-95%.

PB abatement was monitored with a diode array detector (DAD; λ = 276 nm) and a Novapack C_{18} (Waters, USA) column. CH_3CN/H_2O (60/40; v/v) was used as mobile phase for the analysis of PB at a flow rate of 1 mL min^{-1}. The column temperature was set as 30°C for all measurements and the injection volume was selected as 30 µL. The instrument detection limit (1.5 mg L^{-1}) for PB was determined as the lowest injected standard that gave a signal-to-noise ratio of at least 3 and an accuracy of 80-95%. The limit of quantification was calculated as 10 times of the signal-to-noise ratio as 5 mg L^{-1}.

In case of Photo-Fenton Fe^{2+}/H$_2$O$_2$/UV-C experiments, all analytical measurements were performed after filtration of the treated samples though a 0.45 µm membrane filter (Sartorius) in order to remove the settled Fe(OH)$_3$ sludge. The pH was measured with a Thermo Orion model 720 pH-meter at any stage of the experiments.

2.6 Kinetic analysis

DOS and PB abatements during H$_2$O$_2$/UV-C and Photo-Fenton treatment processes followed pseudo-first order kinetics with respect to the model pollutant (DOS and PB). Hence, the kinetic rates could be expressed as follows;

$$\ln C/C_o = - k_C \times t \tag{17}$$

where k_C is the pseudo-first order rate constant (in min^{-1}), t stands for photochemical treatment time (in min), C_o and C are the initial and final concentrations of model pollutant (DOS and PB, in mg L^{-1}).

2.7 Electrical energy requirements of the photochemical treatment processes

The photodegradation of aqueous organic pollutants is an energy-intensive process, thus consideration of figures-of-merit definition based on the electrical energy input is appropriate. In order to compare treatment efficiencies for photochemical oxidation processes, it is important to use a general expression and relate the treatment efficiencies directly to operating costs. For this purpose, the electrical energy per order (EE/O) defined as the kWh electrical energy required to degrade a contaminant or group parameter by one order of magnitude in m^3 of contaminant water or wastewater, can be calculated for treatment process (Bolton et al., 2001). Since the EE/O (kWh m^{-3} order^{-1}) concept assumes pseudo-first order degradation kinetics with respect to pollutant concentration can be determined from the following formula;

$$EE/O \ (kWh \ m^{-3}order^{-1}) = (38.4 \times P) \ / \ (V \times k_C) \tag{18}$$

where P is the lamp power (0.04 kW in the present work), V is the reactor volume (3.25 L in the present work), and k_C is the pseudo-first-order rate constant of DOS and PB degradation (min^{-1}).

3. Results and discussion

3.1 Activated sludge inhibition experiments with untreated DOS and PB

The ISO 8192 (2007) specifies a method for assessing the inhibitory or stimulatory effects of substances, mixtures or wastewaters to activated sludge (Paixão et al., 2002). Information generated by this method may be helpful in estimating the effect of a test material on bacterial communities in the aquatic environment, especially in aerobic biological treatment systems (Paixão et al., 2002). In the first part of this experimental study, untreated (original) aqueous DOS and PB solutions were individually exposed to activated sludge inhibition experiments. The obtained semi-logarithmic I$_{OUR}$ (in %) versus DOS and PB concentrations (log DOS and log PB) were presented in Figure 1 (a) and (b), respectively for an incubation period of 15 min.

Effect of Photochemical Advanced Oxidation Processes on the Bioamenability and Acute
Toxicity of an Anionic Textile Surfactant and a Textile Dye Precursor

89

From Figure 1 (a) it is evident that increasing the DOS concentration resulted in an accelerated inhibition of the OUR value measured relative to the sample blank (synthetic sewage). The I_{OUR} values ranged between 4%-53% from the lowest to highest studied concentrations, respectively, enabling the determination of EC_{30} and EC_{50} values (effective concentrations causing 30 and 50% inhibition in heterotrophic biomass relative to the blank) for DOS, as shown in the insert of Figure 1 (a). These values were established as 338 and 544 mg L^{-1} corresponding to the respirometric inhibition levels of 30% and 50%, respectively. This information implied that DOS is an acutely toxic compound at concentrations encountered in the textile preparation process. Hence its treatment after discharge becomes essential. On the other hand, Figure 1 (b) reveals that, although the studied concentration range for PB is appreciably higher, activated sludge inhibition rates always remained below 20%. From the obtained experimental results it can be concluded that PB is not exhibiting an inhibitory/toxic effect on heterotrophic biomass relative to synthetic sewage. However, it should be noted that the inert COD content of PB creates a significant effluent discharge penalty problem for dye synthesis effluent. For both chemicals it is obvious that efficient treatment employing advanced oxidation process is necessary to comply with the environmental regulations and to minimize pollution loads.

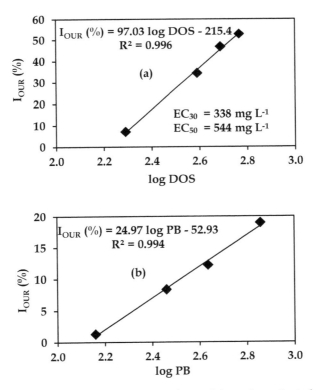

Fig. 1. Log(concentration) versus I_{OUR} (%) curves obtained from the activated sludge inhibition experiments for aqueous DOS (a) and PB (b). DOS concentration range: 97-584 mg L^{-1}; PB concentration range: 144-721 mg L^{-1}. The selected incubation time is 15 min.

3.2 H_2O_2/UV-C and Photo-Fenten treatment of DOS and PB

Figure 2 presents the H_2O_2/UV-C and Photo-Fenton treatment results for aqueous DOS (a, H_2O_2/UV-C; b, Photo-Fenton) and PB (c, H_2O_2/UV-C; d, Photo-Fenton) solutions in terms of DOS-PB, COD and TOC abatement rates as a function of photochemical treatment time. In the Photo-Fenton experiments all experimental conditions were kept identical with the exception of 0.5 mM Fe^{2+} addition to the reaction solution. The initial COD of the DOS-PB solutions were selected as 450 mg L^{-1} in these experiments, which is an average, typical COD for effluents containing these model compounds as the main organic pollutant.

As it can be seen from Figure 2 (a) and (b) complete DOS removal was obtained after 60 min for both treatment processes. In terms of the COD parameter, 84% and 92% removals were achieved after 100 min H_2O_2/UV-C and Photo-Fenton treatment, respectively. TOC removals were also close to another; these were obtained as 89% after 100 min H_2O_2/UV-C treatment and 90% at the end of 100 min Photo-Fenton treatment. From these findings it can be concluded that DOS, COD and TOC removal efficiencies were not significantly different or higher for the Photo-Fenton process as compared to H_2O_2/UV-C treatment. Considering that the addition of Fe^{2+} as a photocatalyst and extra chemical as well as the requirement of acidic reaction conditions for iron-based AOPs will affect the operating costs for DOS treatment considerably, it was decided to keep the H_2O_2/UV-C process for DOS treatment. Kinetic rate coefficients and electrical energy calculations shown in Table 2 and explained later on supported this argument.

Figure 2 (c) and (d) depict the results of H_2O_2/UV-C and Photo-Fenton treatment of aqueous PB solution, respectively. As expected, the addition of 0.5 mM Fe^{2+} had a significant positive effect in terms of PB as well as COD and TOC abatement rates as compared with the H_2O_2/UV-C process (see Figure 2 (c) and (d)). Total PB elimination was realized only after 30 min Photo-Fenton treatment, whereas at least 180 min were needed when PB was subjected to H_2O_2/UV-C treatment. Moreover, COD and TOC were only poorly removed via mere H_2O_2/UV-C treatment. For instance, parallel to the slow COD abatement, TOC practically remained unchanged indicating that PB, the mother compound, was converted to different advanced oxidation intermediates but not ultimately oxidized to mineralization end products in the absence of Fe^{2+} ions. At the end of the 120 min treatment period, COD and TOC removals were obtained as 49% (H_2O_2/UV-C treatment) and 98% (Photo-Fenton treatment) as well as 33% (H_2O_2/UV-C treatment) and 100% (Photo-Fenton treatment), respectively. Considering that H_2O_2/UV-C treatment of PB resulted in very poor and insufficient results, Photo-Fenton oxidation appears by far the better option for the efficient treatment of PB.

The results obtained for DOS and PB could be partially attributed to the high absorbance (optical density) of PB at the maximum emission band of the UV-C light source, hindering its effective absorption by H_2O_2, whereas DOS did not compete with H_2O_2 for UV-C light absorption at 254 nm wavelength that is required for the production of HO• radicals via UV-C.

Table 2 summarizes the calculated k_C and EE/O values for the photochemical treatment of DOS and PB in the absence and presence of Fe^{2+} catalyst. From Table 2 it is obvious that for DOS treatment, Fe^{2+} is not essential, whereas the positive effect of the catalyst was very pronounced for PB abatement. Hence, in the forthcoming experiments where changes in the acute toxicity of DOS and PB during photochemical treatment were examined, results were presented only for the selected AOPs (e.g. H_2O_2/UV-C for DOS and Photo-Fenton for PB model pollutants).

Effect of Photochemical Advanced Oxidation Processes on the Bioamenability and Acute
Toxicity of an Anionic Textile Surfactant and a Textile Dye Precursor

91

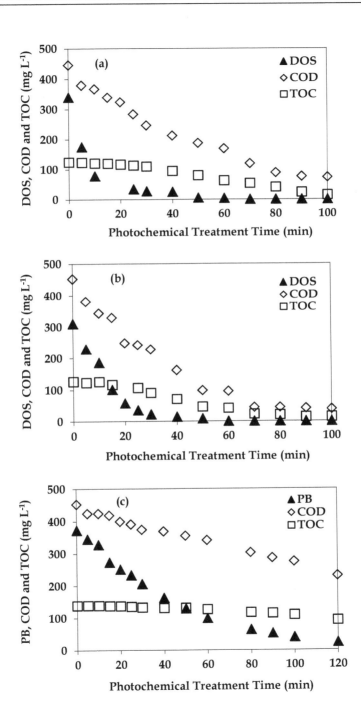

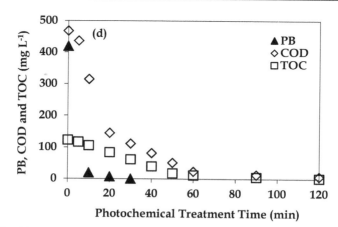

Fig. 2. DOS, PB, COD and TOC abatement rates observed during photochemical treatment of DOS (a, H_2O_2/UV-C; b, Photo-Fenton) and PB (c, H_2O_2/UV-C; d, Photo-Fenton). For H_2O_2/UV-C experiments: Initial COD= 450 mg L^{-1}; H_2O_2 = 30 mM (DOS) and 60 mM (PB); pH = 10.5 (DOS) and 3.0 (PB). For Fe^{2+}/H_2O_2/UV-C experiments: Initial COD= 450 mg L^{-1}; H_2O_2 = 30 mM; Fe^{2+} = 0.5 mM; pH = 3.0.

Model Pollutants	H_2O_2/UV-C Initial COD= 450 mg L^{-1} H_2O_2 = 30 mM (DOS) and 60 mM (PB) pH = 10.5 (DOS) and 3.0 (PB)		Fe^{2+}/H_2O_2/UV-C Initial COD= 450 mg L^{-1} H_2O_2 = 30 mM; Fe^{2+} = 0.5 mM pH = 3.0	
	k_C (min^{-1})	EE/O (kWh m^{-3} order^{-1})	k_C (min^{-1})	EE/O (kWh m^{-3} order^{-1})
DOS	0.1500	3.15	0.1016	4.65
PB	0.0230	20.55	0.2084	2.27

Table 2. First-order abatement rate constants and EE/O values derived for the treatment of aqueous DOS and PB solutions with the H_2O_2/UV-C and Fe^{2+}/H_2O_2/UV-C processes.

3.3 Activated sludge inhibition results for photochemically treated DOS and PB

The possibility exists that during the applicability of AOPs, more toxic and/or less bioamenable intermediates and/or end products form and accumulate in the reaction solution. In order to evaluate the acute toxicity and bioamenability of photochemical degradation products towards heterotrophic biomass, activated sludge inhibition experiments were conducted during H_2O_2/UV-C of DOS and Photo-Fenton oxidation of PB (Figure 3).

For the activated sludge experiments the OUR of the blank samples (containing only synthetic sewage) was adjusted to around 100 mg L^{-1} h^{-1}. The OURs of the sample blanks were also given in Figure 3. Figure 3 indicated that a slight increase in the OUR values is observed during photochemical treatment of DOS starting with 71 mg L^{-1} h^{-1} and reaching its highest value after 40 min photochemical oxidation. Thereafter, the OUR value dropped down even below its original value indicating that extended oxidation did not improve but reduce the quality of DOS bearing effluent most probably due to the formation of less

Effect of Photochemical Advanced Oxidation Processes on the Bioamenability and Acute
Toxicity of an Anionic Textile Surfactant and a Textile Dye Precursor

93

biodegradable and hence more inhibitory end products. The trend observed for the OUR values of the PB samples exposed to Photo-Fenton was different; the OUR values increased throughout the entire treatment period from 98 mg L^{-1} h^{-1} at the beginnig up to 120 mg L^{-1} h^{-1} after 80 min Photo-Fenton treatment. These results demonstrate that PB does not cause any inibitory effect of heterotrophic biomass and the treated reaction solution was even accepted as an extra carbon source for the bioculture as advanced oxidation progressed. The OUR results implied that no acutely toxic/bioinhibitory advanced oxidation products were formed during Photo-Fenton treatment of PB.

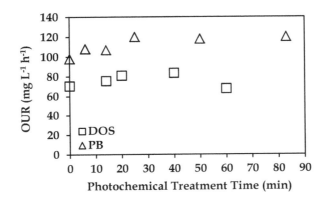

Fig. 3. OUR values obtained during photochemical treatment of DOS via H_2O_2/UV-C (H_2O_2=42 mM; pH=10.5) and PB via Photo-Fenton (Fe^{2+}= 0.72 mM; H_2O_2=40 mM; pH=3.0) processes. Experimental conditions: Initial COD = 450 mg L^{-1}; Incubation period = 15 min; OUR_{blank} for DOS = 92 mg L^{-1} h^{-1}; OUR_{blank} for PB = 118 mg L^{-1} h^{-1}.

4. Conclusion

The present study aimed at exploring the effect of photochemical advanced oxidation processes (H_2O_2/UV-C and Photo-Fenton treatment) on the treatability and bioamenability of two frequently used industrial chemicals; i.e. an anionic textile surfactant (called DOS herein) and a vinylsuphone dye precursor (called PB herein). The following experimental results were obtained in this work;

- The anionic surfactant DOS exhibited a moderate acute toxicity towards heterotrophic biomass; the respirometric inhibition rate varied between 4% to 53% for the lowest and highest studied DOS concentrations, respectively. Accordingly, EC_{30} and EC_{50} values were established as 338 mg L^{-1} and 544 mg L^{-1}, respectively, for the textile preparation surfactant.
- For the textile dye precursor PB, on the other hand, relative activated sludge inhibition rates always remained below 20% and hence no EC values could be calculated for this model pollutant. Conclusively, PB did not exhibit any toxic effect on activated sludge microorganisms in the studied concentration range.
- DOS, a relatively simply structured, aliphatic compound, could be effectively treated via both H_2O_2/UV-C and Photo-Fenton processes. The difference in the overall removal efficiencies and reaction rate coefficient was not significant and hence it appeared that

mere H_2O_2/UV-C treatment in the absence of Fe^{2+} photocatalyst was sufficient for complete DOS and associated organic carbon content abatements. The main advantage of the H_2O_2/UV-C treatment process over Photo-Fenton treatment is its robustness towards reaction pH below pH = 11 hence not requiring a pH adjustment to acidic values.

• For aromatic compounds that strongly absorp UV-C light at 254 nm thus hindering its efficient use by H_2O_2 for HO• production, Photo-Fenton treatment improves the degradation rate considerably. PB abatement via H_2O_2/UV-C treatment process resulted in relatively slow and incomplete degradation rates. In the presence of Fe^{2+}, however, PB degradation was enhanced by a factor of 10 and its removal was complete after 30-40 min Photo-Fenton treatment. These results suggest that the H_2O_2/UV-C process is not capable of effectively degrading the complicated aromatic structure of PB and under these circumstances; Photo-Fenton treatment is the treatment process of choice.

According to the activated sludge inhibition tests no toxic/bioinhibitory degradation intermediates and/or oxidation end products were formed during and after photochemical treatment of DOS (60 min H_2O_2/UV-C treatment at an initial pH of 10.5 with 42 mM H_2O_2) and PB (Photo-Fenton treatment for 80 min at pH 3; in the presence of 0.72 mM Fe^{2+} and 40 mM H_2O_2). The obtained findings indicated that photochemical advanced oxidation processes are potential, promising options for the effective treatment of refractory and toxic industrial pollutants.

5. Acknowledgments

The authors acknowledge the financial support of the Scientific and Technological Research Council of Turkey (TUBITAK) under project number 108Y051 and Eksoy Chemicals for the DOS and PB gift samples. The authors also thank Ms. Zeynep Kartal and Ms. Elif Dedetas for their help in the photochemical treatment experiments.

6. References

American Public Health Association/American Water Works Association/Water Environment Federation (2005). *Standard Methods for the Examination of Water and Wastewater*, 21st Edn., APHA-AWWA-WEF, ISBN 0875530478, Washington D.C., U.S.A.

Arslan-Alaton, I. & Erdinc., E. (2006). Effect of Photochemical Treatment on the Biocompatibility of a Commercial Nonionic Surfactant Used in the Textile Industry. *Water Research*, Vol.40, No.18, (October 2006), pp. 3409-3418, ISSN 0043-1354

Arslan-Alaton, I., Cokgor, E.U. & Koban, B. (2007). Integrated Photochemical and Biological Treatment of a Commercial Textile Surfactant: Process Optimization, Process Kinetics and COD Fractionation. *Journal of Hazardous Materials*, Vol.146, No.3, (July 2007), pp. 453-458, ISSN 0304-3894

Arslan-Alaton, I., Olmez-Hanci, T., Gursoy, B.H. & Tureli, G. (2009a). H_2O_2/UV-C Treatment of the Commercially Important Aryl Sulfonates H-, K-, J-acid and Para Base: Assessment of Photodegradation Kinetics and Products. *Chemosphere*, Vol.76, No.5, (July 2009), pp. 587-594 , ISSN 0045-6535

Effect of Photochemical Advanced Oxidation Processes on the Bioamenability and Acute
Toxicity of an Anionic Textile Surfactant and a Textile Dye Precursor

95

Arslan-Alaton, I., Tureli, G. & Olmez-Hanci, T. (2009b). Treatment of Azo Dye Production Wastewaters using Photo-Fenton-Like Advanced Oxidation Processes: Optimization by Response Surface Methodology. *Journal of Photochemistry and Photobiology A: Chemistry*, Vol.202, No.2-3, (February 2009), pp. 142-153, ISSN 1010-6030

Arslan-Alaton, I., Ayten, N. & Olmez-Hanci, T. (2010). Photo-Fenton-like Treatment of the Commercially Important H-Acid: Process Optimization by Factorial Design and Effects of Photocatalytic Treatment on Activated Sludge Inhibition. *Applied Catalysis B: Environmental*, Vol.96, No.1-2, (April 2010), pp. 208-217, ISSN 0926-3373

Baxendale, J. & Wilson, J. (1957). The Photolysis of Hydrogen Peroxide at High Light Intensities. *Transactions of the Faraday Society*, Vol.53, pp. 344-356, ISSN 0014-7672

Bolton, J.R., Bircher, K.G., Tumas, W. & Tolman, C.A. (2001). Figure-of-Merit for the Technical Development and Application of Advanced Oxidation Technologies for both Electric-and Solar-Driven Systems. *Pure and Applied Chemistry*, Vol.73, No.4, (April 2001), pp. 627-637, ISSN 0033-4545

Calgon Carbon Oxidation Technologies (1995). *The AOP Handbook*, Calgon Carbon Oxidation Technologies (CCOT), Markham, Ontario, Canada

Chen, R. & Pignatello, J. (1997). Role of Quinone Intermediates as Electron Shuttles in Fenton and Photoassisted Fenton Oxidations of Aromatic Compounds. *Environmental Science and Technology*, Vol.31, No.8, (August 1997), pp. 2399-2406, ISSN 0013-936X

Dalzell, D.J.B., Alte, S., Aspichueta, E., de la Sota, A., Etxebarria, J., Gutierrez, M., Hoffmann, C.C., Sales, D., Obst, U. & Christofi, N. (2002). A Comparison of Five Rapid Direct Toxicity Assessment Methods to Determine Toxicity of Pollutants to Activated Sludge. *Chemosphere*, Vol.47, No.5, (May 2002), pp. 535-545, ISSN 0045-6535

European Union (2003). *Reference Document on Best Available Techniques for the Textiles Industry*, European Integrated Pollution Prevention Control (IPPC) Bureau, Seville, Spain.

Farré, M.J., Brosillon, S., Doménech, X. & Peral, J. (2007). Evaluation of the Intermediates Generated during the Degradation of Diuron and Linuron Herbicides by the Photo-Fenton Reaction, *Journal of Photochemistry and Photobiology A: Chemistry*, Vol.188, No.2-3, (June 2007), pp. 34-42, ISSN 1010-6030

Gendig, C., Domogala, G., Agnoli, F., Pagga, U. & Strotmann, U.J. (2003). Evaluation and Further Development of the Activated Sludge Respiration Inhibition Test. *Chemosphere*, Vol.52, No.1, (July 2003), pp. 143-149, ISSN 0045-6535

González, O., Sans, C. & Esplugas, S. (2007). Sulfamethoxazole Abatement by Photo-Fenton. Toxicity, Inhibition and Biodegradability Assessment of Intermediates. *Journal of Hazardous Materials*, Vol.146, No.3, (July 2007), pp. 459-464, ISSN 0304-3894

Henriques, I.D.S., Holbrook, R.D., Kelly, R.T. & Love, N.G. (2005). The Impact of Floc Size on Respiration Inhibition by Soluble Toxicants-A Comparative Investigation. *Water Research*, Vol.39, No.12, (July 2005), pp. 2559-2568, ISSN 0043-1354

Ikehata, K. & El-Din, M.G. (2004). Degradation of Recalcitrant Surfactants in Wastewater by Ozonation and Advanced Oxidation Processes: A Review. *Ozone Science and Engineering*, Vol.26, No.4, (August 2004), pp. 327-343, ISSN 0191-9512

International Standardization Organization (1989). *Water Quality-Determination of the Chemical Oxygen Demand*, 2nd Edn., ISO 6060/TC 147, ICS 13.060.50, Geneva, Switzerland

International Standardization Organization (2007). *Water Quality-Test for Inhibition of Oxygen Consumption by Activated Sludge for Carbonaceous and Ammonium Oxidation*, 2nd Edn., ISO 8197/TC 147, ICS 13.060.70, Geneva, Switzerland

Jandera, P., Fischer, J. & Prokes, B. (2001). HPLC Determination of Chlorobenzenes, Benzenesulfonyl Chlorides and Benzenesulfonic Acids in Industrial Wastewater. *Chromatographia*, Vol.54, No.9-10, (November 2001), pp. 581-587, ISSN 0009-5893

Marco, A., Esplugas, S. & Saum, G. (1997). How and Why Combine Chemical and Biological Processes for Wastewater Treatment. *Water Science and Technology*, Vol.35, No.4, pp. 321-327, ISSN 0273-1223

Muruganandham, M. & Swaminathan, M. (2004). Decolourisation of Reactive Orange 4 by Fenton and Photo-Fenton Oxidation Technology. *Dyes and Pigments*, Vol.63, No.3, (December 2004), pp. 315-321, ISSN 0143-7208

Nicole, I., De Laat, J., Dore, M., Duguet, J. P. & Bonnel, C. (1990). Utilisation du Rayonnement Ultraviolet dans le Traitement de Waux: Measure du Flux Photonique par Actinometrie Chimique au Peroxide d'Hydrogene. *Water Research*, Vol.24, No.2, (February 1990), pp. 157-168, ISSN 0043-1354

Official Methods of Analysis (1980). *Association of Official Analytical Chemists (AOAC International)*, ISBN 0935584145, Washington D.C., U.S.A.

Oliveros, E., Legrini, O., Hohl, M., Müller, T. & Braun, A.M. (1997). Industrial Wastewater Treatment: Large-Scale Development of a Light-Enhanced Fenton Reaction. *Chemical Engineering and Processing*, Vol.36, No.5, (September 1997), pp. 397-405, ISSN 0255-2701

Oller, I., Malato, S., Sánchez-Pérez, J.A., Maldonado, M.I. & Gassó, R. (2007). Detoxification of Wastewater Containing Five Common Pesticides by Solar AOPs-Biological coupled System, *Catalysis Today*, Vol.129, No.1-2, (November 2007), pp. 69-78, ISSN 0920-5861

Olmez-Hanci, T., Imren, C., Arslan-Alaton, I., Kabdaşlı, I. & Tünay, O. (2009). H_2O_2/UV-C Oxidation of Potential Endocrine Disrupting Compounds: A Case Study with Dimethyl Phthalate. *Photochemical and Photobiological Sciences*, Vol.8, No.5, pp. 620-627, ISSN 1474-905X

Oppenländer, T. (2003). *Photochemical Purification of Water and Air: Advanced Oxidation Processes (AOPs): Principles, Reaction Mechanisms, Reactor Concepts*, John Wiley, Inc., ISBN 3527305637, Chichester, U.K.

Paixão, S.M. & Anselmo, A.M. (2002). Effect of Olive Mill Wastewaters on the Oxygen Consumption by Activated Sludge Microorganisms: An Acute Toxicity Test Method. *Journal of Applied Toxicology*, Vol.22, No.3, (May-June 2002), pp. 173-176, ISSN 0260-437X

Petrovic, M. & Barceló, D. (2004). Fate and Removal of Surfactants and Related Compounds in Wastewater and Sludges, In: *The Handbook of Environmental Chemistry Part I.*, Vol. 5, 95-108, Springer, ISSN 1867-979X, Weinheim, Germany

Pignatello, J.J. (1992). Dark and Photo-Assisted Fe^{3+}-Catalyzed Degradation of
    Chlorophenoxy Herbicides by Hydrogen Peroxide. *Environmental Science and
    Technology*, Vol.26, No.5, (May 1992), pp. 944-951, ISSN 0013-936X
Prousek, J. (1996). Advanced Oxidation Processes for Water Treatment. Chemical Processes,
    Chemical Listy, Vol.90, No.4, (April 1996), pp. 229-237, ISSN 0009-2770
Rodríguez, M. (2003). *Fenton and UV-Vis Based Advanced Oxidation Processes in Wastewater
    Treatment: Degradation, Mineralization and Biodegradability Enhancement*, PhD Thesis,
    Universitat De Barcelona, Facultat De Química, Departament D'enginyeria
    Química I Metal ·Lúrgia, Barcelona, Spain
Sagawe, G., Lehnard, A., Lubber, M., Bahnemann, D. (2001). The Insulated Solar Fenton
    Hybrid Process: Fundamental Investigations, *Helvetica Chimica Acta*, Vol.84, No.12,
    pp. 3742-3759, ISSN 0018-019X
Scott, J.P. & Ollis, D.F. (1995). Integration of Chemical and Biological Oxidation Processes for
    Water Treatment: Review and Recommendations. *Environmental Progress*, Vol.14,
    No.2, (May 1995), pp. 88-98, ISSN 0278-4491
Staples, C.A., Naylor, C.G., Williams, J.B. & Gledhill, W.E. (2001). Ultimate Biodegradation
    of Alkylphenol Ethoxylate Surfactants and their Biodegradation Intermediates,
    Environmental Toxicology and Chemistry, Vol.20, No.11, (November 2001), pp. 2450-
    2455, ISSN 0730-7268
Stasinakis, A.S., Petalas, A.V., Mamais, D. & Thomaidis, N.S. (2008). Application of the
    OECD 301F Respirometric Test for the Biodegradability Assessment of Various
    Potential Endocrine Disrupting Chemicals. *Bioresources Technology*, Vol.99, No.9,
    (June 2008), pp. 3458-3467, ISSN 0960-8524
Sychev, A.Y. & Isak, V.G. (1995). Iron Compounds and the Mechanisms of the
    Homogeneous Catalysis of the Activation of O_2 and H_2O_2 and of the Activation of
    Organic Substrates. *Russian Chemical Reviews*, Vol. 64, No.12, pp. 1105-1129, ISSN
    0036-021X
Swisher, R.D. (1987). *Surfactant Biodegradation*, 2nd Edn., Marcel Dekker, ISBN 0824769384,
    New York, U.S.A.
The European Apparel and Textile Confederation (2000), *Textile Industry BREF Document*,
    Chapters 2-6, Euratex, Brussels, Belgium
Ubay Cokgor, E., Ozdemir, S., Karahan, O., Insel, G. & Orhon, D. (2007). Critical Appraisal
    of Respirometric Methods for Metal Inhibition on Activated Sludge. *Journal of
    Hazardous Materials B*, Vol.139, No.2, (January 2007), pp. 332-339, ISSN 0304-3894
Utsunomiya, A., Watanukii, T., Matsusi-Fita, K., Nishina, M. & Tomita, I. (1997). Assessment
    of the Toxicity of Linear Alkylbenzene Sulfonate and Quaternary Alkylammonium
    Chloride by measuring 13C-Glycerol in *Dunaliella sp. Chemosphere*, Vol.35, No.11,
    (December 1997), pp. 2479-2490, ISSN 0045-6535
Van de Plassche, E.J., De Bruijn, J.H.M., Stephenson, R.R., Marshall, S.J., Feijtel, T.C.J. &
    Belanger, S.E. (1999). Predicted No-Effect Concentrations and Risk Characterization
    of Four Surfactants: Linear Alkyl Benzene Sulfonate, Alcohol Ethoxylates, Alcohol
    Ethoxylated Sulfates, and Soap. *Environmental Toxicology and Chemistry*, Vol.18,
    No.11, (November 1999), pp. 2653-2663, ISSN 0730-7268

Wadley, S. & Waite, T.D. (2004). Fenton Processes, In: *Advanced Oxidation Processes for Water and Wastewater Treatment*, S. Parsons, (Ed.), 111-136, IWA Publishing, ISBN 1843390175, Padstow, Cornwall, U.K.

Wang, C. X., Yediler, A., Lienert, D., Wang, Z. & Kettrup, A. (2002). Toxicity Evaluation of Reactive Dyestuff, Auxiliaries and Selected Effluents in Textile Finishing Industry to Luminescent Bacteria *Vibrio fischeri*. *Chemosphere,* Vol.46. No.2, (January 2002), pp. 339-344, ISSN 0045-6535

Zollinger, H. (2001). *Color Chemistry*, 3rd Edn., Wiley-VCH, ISBN 3906390233, Weinheim, Germany

5

Azo Dyes and Their Metabolites: Does the Discharge of the Azo Dye into Water Bodies Represent Human and Ecological Risks?

Farah Maria Drumond Chequer[1],
Daniel Junqueira Dorta[2] and Danielle Palma de Oliveira[1]
[1]*USP, Departamento de Análises Clínicas, Toxicológicas e Bromatológicas,*
Faculdade de Ciências Farmacêuticas de Ribeirão Preto,
Universidade de São Paulo, Ribeirão Preto – SP,
[2]*USP, Departamento de Química,*
Faculdade de Filosofia, Ciências e Letras de Ribeirão Preto,
Universidade de São Paulo, Ribeirão Preto – SP,
Brazil

1. Introduction

1.1 History of sintetic dyes

Colorants (dyes and pigments) are important industrial chemicals. According to the technological nomenclature, pigments are colorants which are insoluble in the medium to which they are added, whereas dyes are soluble in the medium. The world's first commercially successful synthetic dye, named mauveine, was discovered by accident in 1856 by William Henry Perkin. These synthetic compounds can be defined as colored matters that color fibers permanently, such that they will not lose this color when exposed to sweat, light, water and many chemical substances including oxidizing agents and also to microbial attack (Rai et al., 2005; Saratele et al., 2011). By the end of the 19th century, over ten thousand synthetic dyes had been developed and used for manufacturing purposes (Robinson et al., 2001a; Saratele et al., 2011), and an estimate was made in 1977 that approximately 800,000 tons of all recognized dyestuffs had been produced throughout the world (Anliker, 1977; Combes & Haveland-Smith, 1982). The expansion of worldwide textile industry has led to an equivalent expansion in the use of such synthetic dyestuffs, resulting in a rise in environmental pollution due to the contamination of wastewater with these dyestuffs (Pandey et al., 2007; Saratele et al., 2011).

The Ecological and Toxicological Association of the Dyestuffs Manufacturing Industry (ETAD) was inaugurated in 1974 with the goals of minimizing environmental damage, protecting users and consumers and cooperating with government and public concerns in relation to the toxicological impact of their products (Anliker, 1979; Robinson et al., 2001a). A survey carried out by ETAD showed that of a total of approximately 4,000 dyes that had been tested, more than 90% showed LD_{50} values above 2×10^3 mg/kg, the most toxic being in the group of basic and direct diazo dyes (Shore, 1996; Robinson et al., 2001a). Thus it appears that exposure to azo dyes does not cause acute toxicity, but with respect to systemic bioavailability, inhalation and contact with the skin by azo dyes is of concern, due to the possible generation of carcinogenic aromatic amines (Myslak & Bolt, 1988 and Bolt & Golka, 1993 as cited in Golka et al., 2004).

Of the approximately 10^9 kg of dyestuffs estimated to be manufactured annually throughout the World, the two most widely used in the textile industry are the azo and anthraquinone groups (Križanec & Marechal, 2006; Forss, 2011). Thus, this chapter is a comprehensive review on the azo dyes and their effects on human and environmental health.

2. Azo dyes

Azo dyes are diazotized amines coupled to an amine or phenol, with one or more azo bonds (–N=N–). They are synthetic compounds and account for more than 50% of all the dyes produced annually, showing the largest spectrum of colors (Carliell et al., 1995; Bae & Freeman, 2007; Kusic et al., 2011). Nearly all the dyestuffs used by the textile industry are azo dyes, and they are also widely used in the printing, food, papermaking and cosmetic industries (Chung & Stevens, 1993; Chang et al., 2001a). An estimate was made in the 80's, that 280,000 t of textile dyes were annually discharged into industrial effluents worldwide (Jin et al., 2007; Saratale et al., 2011). Since the azo dyes represent about 70% by weight of the dyestuffs used (Zollinger, 1987), it follows that they are the most common group of synthetic colorants released into the environment (Chang et al., 2001b; Zhao & Hardin, 2007; Saratale et al., 2011).

One only needs very small amounts of dyes in the water (less than 1 ppm for some dyes) to cause a highly visible change in color (Banat et al., 1996), and colored wastewater not only affects the aesthetic and transparency aspects of the water being received, but also involves possible environmental concerns about the toxic, carcinogenic and mutagenic effects of some azo dyes (Spadaro et al., 1992; Modi et al., 2010; Lu et al., 2010). It can also affect the aquatic ecosystem, decreasing the passage of light penetration and gas dissolution in lakes, rivers and other bodies of water (Saranaik & Kanekar, 1995; Banat et al., 1996; Modi et al., 2010).

The more industrialized the society, the greater the use of azo dyes, and hence the greater the risk of their toxic effects affecting the society. It has already been noted that, as from the 70's, intestinal cancer has been more common in highly industrialized societies, and therefore there may be a connection between the increase in the number of cases of this disease and the use of azo dyes (Wolff & Oehme, 1974; Chung et al., 1978).

Bae and Freeman (2007) already demonstrated the biological toxicity of the direct azo dyes used in the textile industry. The results indicated that C.I. Direct Blue 218 was very toxic to daphnids, with a 48-h LC_{50} between 1.0 and 10.0 mg/L. It must be remembered that toxicity to daphnids is sufficient to suggest potential damage to every receptor ecosystem, and emphasizes the need for the synthetic dye manufacturing industry to carry out toxicological studies (Bae & Freeman, 2007).

2.1 Azo dyes and their mutagenic effects

The azo dyes show good fiber-fixation properties as compared other synthetic dyes, showing up to 85% fixation, but nevertheless this explains why so much dye is released into

Azo Dyes and Their Metabolites: Does the Discharge of the Azo Dye into Water
Bodies Represent Human and Ecological Risks?

101

the environment, representing the other 10 to 15% of the amount used. Most of the synthetic dyestuffs found in this class are not degraded by the conventional treatments given to industrial effluents or to the raw water (Nam & Reganathan, 2000; Oliveira et al., 2007). Shaul et al. (1991) studied 18 azo dyes, and found that 11 passed practically unchanged through the activated sludge system, 4 were adsorbed by the activated sludge and only 3 were biodegraded, resulting in the release of these substances into bodies of water. Oliveira et al. (2007) showed that even after treatment, effluent from dyeing industries was mutagenic and contained various types of dye. Such data are of concern, especially when one considers that the effluent from the same industry was studied by Lima et al. (2007), who found an increase in the incidence of aberrant crypt in the colon of rats exposed to this sample, this being an early biomarker of carcinogenesis (Lima et al., 2007).

Azo dyes can also be absorbed after skin exposure, and such dermal exposure to azo dyes can occur as an occupational hazard or from the use of cosmetic products. It was postulated in the 80s that the percutaneous absorption of azo dyes from facial makeup could even be a risk factor in reproductive failures and chromosomal aberrations in a population of television announcers (Kučerová et al., 1987; Collier et al., 1993).

Various azo dyes have been shown to produce positive toxic results for different parameters. Tsuboy et al. (2007) analyzed the mutagenic, cytotoxic and genotoxic effects of the azo dye CI Disperse Blue 291, and the results clearly showed that this azo dye caused dose-dependent effects, inducing the formation of micronuclei (MNs), DNA fragmentation and increasing the apoptotic index in human hepatoma cells (HepG$_2$). A variety of azo dyes have shown mutagenic responses in Salmonella and mammalian assay systems, and it is apparent that their potencies depend on the nature and position of the aromatic rings and the amino nitrogen atom. For instance, 2-methoxy-4-aminoazobenzene is an extremely weak mutagen, whereas under similar conditions, 3-methoxy-4-aminoazobenzene is a potent hepatocarcinogen in rats and a strong mutagen in *Escherichia coli* and *Salmonella typhimurium* (Hashimoto et al., 1977; Esancy et al., 1990; Garg et al., 2002, Umbuzeiro et al., 2005a).

According to Chequer et al. (2009), the azo dyes Disperse Red 1 and Disperse Orange 1 increase the frequency of MNs in human lymphocytes and in HepG2 cells in a dose-dependent manner. According to Ferraz et al. (2010), the azo dyes Disperse Red 1 and Disperse Red 13 showed mutagenic activity in the Salmonella/microsome assay with all the strains tested and in the absence of metabolic activation, except for Disperse Red 13, which was negative with respect to strain TA100. After adding the S9 mix, the mutagenicity of the two azo dyes decreased (or was eliminated), indicating that the P450-dependent metabolism probably generated more stable products, less likely to interact with DNA. It was also shown that the presence of a chlorine substituent in Disperse Red 13 decreased its mutagenicity by a factor of about 14 when compared with Disperse Red 1, which shows the same structure as Disperse Red 13, but without the chlorine substituent. The presence of this substituent did not cause cytotoxicity in HepG2 cells, but toxicity to the water flea *Daphnia similis* increased in the presence of the chlorine substituent (Ferraz et al., 2010).

Chung and Cerneglia (1992) published a review of several azo dyes that had already been evaluated by the Salmonella / microsome assay. According to these authors, all the azo dyes evaluated that contained the nitro group showed mutagenic activity. The dyes Acid Alizarin Yellow R and Acid Alizarin GG showed this effect in the absence of metabolic activation (Brown et al., 1978). The dyes C.I. Basic Red 18 and Orasol Navy 2RB, which also contained nitro groups, were shown to be mutagenic both in the presence and absence of metabolic activation (Venturini & Tamaro, 1979; Nestmann et al., 1981). This review also showed the results obtained in the Salmonella/microsomal test of azo dyes containing benzeneamines, and found that Chrysodin was mutagenic in the presence of a rat-liver preparation (Sole & Chipman, 1986; Chung & Cerneglia, 1992).

Another study applied the micronucleus assay in mouse bone marrow to the azo dye Direct Red 2 (DR2) and the results identified DR2 as a potent clastogen and concluded that excessive exposure to this chemical or to its metabolites could be a risk to human health (Rajaguru et al., 1999).

Al-Sabti (2000) studied the genotoxic effects of exposing the Prussian carp (*Carassius auratus gibelio*) to the textile dye Chlorotriazine Reactive Azo Red 120, and showed its mutagenic activity in inducing MNs in the erythrocytes. They also showed that the dye had clastogenic activity, a potent risk factor for the development of genetic, teratogenic or carcinogenic diseases in fish populations, which could have disastrous effects on the aquatic ecosystem since the fate of compounds found in effluents is to be discharged into water resources (Al-Sabti, 2000).

In addition to the effects caused by exposure to contaminated water and food, workers who deal with these dyes can be exposed to them in their place of work, and suffer dermal absorption. Similarly, if dye-containing effluents enter the water supply, possibly by contamination of the ground water, the general population may be exposed to the dyes via the oral route. This latter point could be of great importance in places where the existent waste treatment systems are inefficient or where there is poor statutory regulation concerning industrial waste disposal (Rajaguru et al., 1999).

2.2 Effects of the azo dyes metabolites

Sisley and Porscher carried out the earliest studies on the metabolism of azo compounds in mammals in 1911, and found sulphanilic acid in the urine of dogs fed with Orange I, demonstrating for the first time that azo compounds could be metabolized by reductive cleavage of the azo group (Sisley & Porscher, 1911 as cited in Walker, 1970).

The mutagenic, carcinogenic and toxic effects of the azo dyes can be a result of direct action by the compound itself, or the formation of free radicals and aryl amine derivatives generated during the reductive biotransformation of the azo bond (Chung et al., 1992; Collier et al., 1993; Rajaguru et al., 1999) or even caused by products obtained after oxidation via cytochrome P450 (Fujita & Peisach, 1978; Arlt et al., 2002; Umbuzeiro et al., 2005a).

One of the criteria used to classify a dye as harmful to humans is its ability to cleave reductively, and consequently generate aromatic amines when in contact with sweat, saliva or gastric juices (Pielesz et al., 1999, 2002). Some such aromatic amines are carcinogenic and can accumulate in food chains, for example the biphenylamines such as benzidine and 4-biphenylamine, which are present in the environment and constitute a threat to human health and to the ecosystems in general (Choudhary, 1996; Chung et al., 2000).

After an azo dye is orally ingested, it can be reduced to free aromatic amines by anaerobic intestinal microflora and possibly by mammalian azo reductase in the intestinal wall or the liver (Walker, 1970; Prival & Mitchel, 1982; Umbuzeiro et al., 2005a). Such biotransformations can occur in a wide variety of mammalian species, including both *Rhesus* monkeys and humans (Rinde & Troll, 1975; Watabe et al., 1980; Prival & Mitchel, 1982;). As

Azo Dyes and Their Metabolites: Does the Discharge of the Azo Dye into Water
Bodies Represent Human and Ecological Risks?

103

previously mentioned, the main biotransformation products of azo dyes are aromatic amines, and thus a brief description of this class of compounds is shown below.

2.2.1 Aromatic amines

As early as the late nineteenth century, a doctor related the occurrence of urinary bladder cancer to the occupation of his patients, thus demonstrating concern about the exposure of humans to carcinogenic aromatic amines produced in the dye manufacturing industry, since his patients were employed in such an industry and were chronically exposed to large amounts of intermediate arylamines. Laboratory investigations subsequently showed that rats and mice exposed to specific azo dye arylamines or their derivatives developed cancer, mainly in the liver (Weisburger, 1997, 2002). Briefly, as mentioned above, in 1895, Rehn showed concern about the urinary bladder cancers observed in three workers from an 'aniline dye' factory in Germany. This led to the subsequent testing in animals of various chemicals to which these workers were exposed, and, as a result, the carcinogenic activity of the azo dye, 2,3-dimethyl-4-aminoazobenzene for the livers of rats and mice was discovered (Yoshida, 1933 as cited in Dipple et al., 1985). An isomeric compound, N,N-dimethyl-4-aminoazobenzene was also found to be a liver carcinogen (Kinosita, 1936 as cited in Dipple et al., 1985). Only in 1954 was the cause of the bladder tumors observed in the workers in the dye industry established to be 2-naphthylamine. This aromatic amine induced bladder cancer in dogs, but not in rats (Hueper et al., 1938 as cited in Dipple et al., 1985).

In addition, workers in textile dyeing, paper printing and leather finishing industries, exposed to benzidine based dyes such as Direct Black 38, showed a higher incidence of urinary bladder cancer (Meal et al., 1981; Cerniglia et al., 1986). Cerniglia et al. (1986) demonstrated that the initial reduction of benzidine-based azo dyes was the result of azoreductase activity by the intestinal flora, and the metabolites of Direct Black 38 were identified as benzidine, 4-aminobiphenyl, monoacetylbenzidine, and acetylaminobiphenyl (Manning et al., 1985; Cerniglia et al., 1986) . Furthermore, these metabolites tested positive in the Salmonella/microsome mutagenicity assay in the presence of S9 (Cerniglia et al., 1986).

In the opinion of Ekici et al. (2001), although general considerations concerning the kinetics of azo dye metabolism indicate that an accumulation of intermediate amines is not very likely, this possibility cannot be excluded under all conditions. According to legislation passed in the European Community on 17th July 1994, the application of azo dyes in textiles is restricted to those colorants which cannot, under any circumstances, be converted to any of the following products: 4-Aminodiphenyl; 4-Amino-2′,3-dimethylazobenzene (*o*-aminoazo-toluene); 4-Aminophenylether (4,4′-oxydianiline); 4-Aminophenylthioether (4,4′-thiodianiline); Benzidine; Bis-(4-aminophenyl)-methane (4,4′-diaminodiphenylme- thane); 4-chloroaniline (*p*-chloroaniline); 4-Chloro-2-methylaniline (4-chloro-*o*-toluidine); 2,4-Diaminotoluene (2,4-toluylenediamine); 3,3′-Dichlorobenzidine dihydrochloride; 3,3′-Dimethoxybenzidine (*o*-dianisidine); 3,3′-Dimethylbenzidine (*o*-toluidine); 3,3′-Dimethyl-4,4′-diamino-diphenyl methane; 2-Methoxy-5-methylaniline (*p*-kresidine); 4-Methoxy-1,3-phenylenediamine sulfate hydrate (2,4-diaminoanisole); 4,4′-Methylene-bis (2-chloroaniline); 2-Methyl-5-nitroaniline (2-amino-4-nitrotoluene); 2-Naphthylamine; *o*-Toluidine; 2,4,5-Trimethylaniline (Bundesgesetzblatt, 1994 and Directory of Environmental Standards, 1998 as cited in Ekici et al., 2001).

More recently, the scientific community has come to consider the possibility of manufactured azo dyes breaking down generating amines to be a health hazard. The International Agency for Research on Cancer only includes benzidine-based dyes in Group 2A and eight other dyes in Group 2B. Nevertheless, the possibility of azo bond reduction leading to the production of aromatic amines has been demonstrated under a variety of conditions, including those encountered in the digestive tract of mammals (Chung & Cerneglia, 1992; Pinheiro et al., 2004). Therefore, the majority of the attention concerning possible hazards arising from the use of azo dyes is now being directed at their reduction products (Pinheiro et al., 2004).

Nitroanilines are aromatic amines that are commonly generated during the biodegradation of azo dyes under anaerobic conditions, formed by reductive cleavage of the azo bonds (–N=N–) by the action of microorganisms present in the wastewaters (Pinheiro et al., 2004; Van der Zee & Villaverde, 2005; Khalid et al., 2009). Depending on the individual compounds, many aromatic amine metabolites are considered to be non-biodegradable or only very slowly degradable (Saupe, 1999), showing a wide range of toxic effects on aquatic life and higher organisms (Weisburger, 2002; Pinheiro et al., 2004; Khalid et al., 2009).

2.3 Metabolic pathways involved in the reduction and oxidation of azo dyes

Following oral or skin exposure to azo dyes, humans can subsequently be exposed to biotransformation products obtained by the action of intestinal microorganisms or that of others present on the skin, or due to reactions in the liver (Esancy et al., 1990; Chadwick et al., 1992; Chung et al., 1992; Stahlmann et al., 2006). Therefore it is extremely important to study the metabolic pathways of azo dyes that can contaminate the environment, in order to understand the overall spectrum of the toxic effects.

The metabolic pathways the azo dyes actually follow depend on several factors, such as, (a) the mode of administration; (b) the degree of absorption from the gastro-intestinal tract after oral ingestion; (c) the extent of biliary excretion, particularly after exposure to different routes other than the oral one; (d) genetic differences in the occurrence and activity of hepatic reducing-enzyme systems; (e) differences in the intestinal flora; and (f) the relative activity and specificity of the hepatic and intestinal systems, particularly those responsible for reducing the azo link, and all these factors are interrelated (Walker, 1970).

Azo dyes behave as xenobiotics, and hence after absorption, they are distributed throughout the body, where they either exert some kind of action themselves or are subjected to metabolism. Biotransformation may produce less harmful compounds, but it may also form bioactive xenobiotics, ie, compounds showing greater toxicity (Kleinow et al., 1987; Livingstone, 1998). The main routes involved in the biotransformation of dyes are oxidation, reduction, hydrolysis and conjugation, which are catalyzed by enzymes (Zollinger, 1991; Hunger, 1994), but in humans, biological reductions and oxidations of azo dyes are responsible for the possible presence of toxic amines in the organism (Pielesz et al., 2002).

Orange II can be reductively metabolized producing 1-amino-2-naphthol, a bladder carcinogen for rats (Bonser et al., 1963; Chung et al., 1992). This suggests that any toxicity induced by unchanged azo dye molecules should not be accepted as the only effect of these compounds, since the reductive cleavage products from these dyes can be mutagenic/carcinogenic (Field et al., 1977; Chung et al., 1992).

Azo Dyes and Their Metabolites: Does the Discharge of the Azo Dye into Water
Bodies Represent Human and Ecological Risks?

105

2.3.1 Oxidative metabolism

Highly lipid-soluble dyes such as azo dyes, with chemical structures containing amino groups, either alkylamino or acetylamino, but without sulfonated groups, are preferentially biotransformed by oxidative reactions (Hunger, 1994).

Oxidation processes are mainly catalyzed by a microsomal monooxygenase system represented by cytochrome P450 (Hunger, 1994), which belongs to a superfamily of heme proteins, present in all living organisms and involved in the metabolism of a wide variety of chemical compounds (Denisov et al. 2005; Mansuy, 2007).

The general mechanism of metabolic oxidation involves an electron transport chain, which first transfers an electron to the P-450-Fe^{3+} complex, which, on reduction, receives an oxygen atom and in the final steps, leads to the formation of an oxidation product in the organism (Furhmann, 1994 as cited in Hunger, 1994).

There are three different oxidation pathways of importance for azo dyes: I) C-Hydroxylation, ring hydroxylation in the case of azo dyes, probably via an epoxidation mechanism and subsequent rearrangement to a phenol. II) N-Hydroxylation at primary or secondary amino groups, or with acetyl amino groups in the liver. This reaction is followed by esterification with glucuronate or sulfate. The activated esters, which are water-soluble, can be excreted, or the ester group can split off with the formation of a nitrenium compound -NH^+, which can covalently bind to a nucleophilic group of the DNA. III) Demethylation, which is the stepwise oxidation of the methyl groups of dialkylamino compounds, and the N-hydroxy derivative so formed can be further demethylated or react to form a nitrenium compound (Hunger, 1994).

Studies on the metabolism and carcinogenicity of N,N-dimethylaminoazobenzene (Butter Yellow), a classical hepatocarcinogen in rats, have shown that N-methylaminoazobenzenes are mainly metabolized by N-demethylation. In this way, Butter Yellow was first reversibly demethylated to the mono-N-methyl compound, which, in turn, was irreversibly demethylated to form p-aminoazobenzene. These changes were shown to precede reduction of the azo link, by isolating N-methyl-p-amino azobenzene and p-aminoazobenzene from the animal tissues (Miller et al., 1945; Walker, 1970). Radiotracer studies have shown (Miller et al., 1952) that demethylation occurs via the formation of a hydroxymethyl compound, followed by elimination of the methyl group in the form of formaldehyde (Mueller & Miller, 1953; Walker, 1970).

Hydroxylation of the aromatic ring can occur before reductive fission of the azo group, and also on the amines produced by such a reduction, and this pathway appears to be very important in compounds which contain an unsulphonated phenyl moiety (Walker, 1970).

2.3.2 Reductive metabolism

Trypan Blue has been shown to have carcinogenic and teratogenic properties (Field et al., 1977). Although original Trypan Blue is not mutagenic, it was reduced by the cell-free extract of an intestinal anaerobe, *Fusobacterium* sp.2, to a mutagenic product, O-toluidine (3,3'-dimethylbenzidine) (Hartman et al., 1978; Chung et al., 1992). In addition to Trypan Blue, Benzopurpurine 4B and Chlorazol Violet N were also shown to be Ames-positive frame-shift mutagens, but only in the presence of metabolizing systems capable of effecting azo reduction. The activity of these dyes may therefore be attributed to the benzidine metabolite, O-toluidine, which is generated because these amines are themselves indirect frame-shift agents (Hartman, et al., 1978; Matsushima et al., 1978). As mentioned above, the benzidine produced after the reduction of some dyes can induce bladder cancer in humans and tumors in some experimental animals (Combes & Haveland-Smith, 1982; Chung, 1983). Some azo dyes, such as Brown FK, have been shown to be directly mutagenic in bacterial tests (Haveland-Smith & Combes, 1980a, b; Rafii et al., 1997). However many other azo dyes, such as Congo Red and Direct Black 38, only give a positive result for mutagenicity after chemical reduction or incubation with the contents of the human intestinal tract (Haveland-Smith & Combes, 1980 a,b; Reid et al., 1983; Cerniglia et al., 1986;;Chung and Cerniglia, 1992; Rafii et al., 1997).

Reductive cleavage of the azo linkages is probably the most toxicologically important metabolic reaction of azo compounds. This reaction can be catalyzed by mammalian enzymes, especially in the liver (Walker, 1970; Kennelly et al., 1982) or by intestinal (Chung et al., 1978; Hartman et al., 1978) or skin bacteria such as Staphylococcus aureus (De France, 1986; Platzek et al., 1999; Golka et al., 2004). Azo compounds can reach the intestine directly after oral ingestion or via the bile after parenteral administration. They are reduced by azo reductases produced by intestinal bacteria, and to a lesser extent by enzymes from the cytosolic and microsomal fractions of the liver. The first catabolic step in the reduction of azo dyes is the cleavage of the azo bond, producing aromatic amines (Cerniglia et al., 1986), accompanied by a loss of color of the dye, and bacterial azoreductases show much greater activity than hepatic azoreductases (Watabe et al., 1980; Collier et al., 1993; Raffi et al., 1997). This reduction process may produce compounds that are more or less toxic than the original molecule (Collier et al., 1993; Rafii et al., 1997), depending on the chemical structure of the metabolite generated. Although its occurrence in the liver has been regarded as the result of a detoxification reaction, azo reduction may be the first step in azo dye carcinogenesis (Chung et al., 1992).

In addition, Nam & Reganathan (2000) demonstrated that both nicotinamide adenine dinucleotide (NADH) and nicotinamide adenine dinucleotide phosphate (NADPH) are capable of reducing azo dyes in the absence of any enzyme, under mildly acidic conditions. The reduced forms of NADH and NADPH are ubiquitous sources of electrons in biological systems, and these function as cofactors in many reductive enzyme reactions. This suggests that the introduction of methyl and methoxy substituents at the 2-,2,3-,2,6-, or 2,3,6-positions of the aromatic ring, accelerates the reduction of phenolic azo dyes by NADH, as compared to that of unsubstituted dyes (Nam & Reganathan, 2000).

It is possible that both the mutagenicity and carcinogenicity of azo dyes are in fact frequently due to the generation of aromatic amines, with subsequent N- and ring hydroxylation and N-acetylation of the aromatic amine (Chung & Cerniglia, 1992). If the azo dyes contain nitro groups, they can also be metabolized by the nitroreductases produced by microorganisms (Chadwick et al., 1992; Umbuzeiro et al., 2005a). Mammalian enzymes in the liver and in other organs can also catalyze the reductive cleavage of the azo bond and the nitroreduction of the nitro group. However, it has been shown that the intestinal microbial azoreductase and nitroreductase play a more important role in this type of metabolism. In both cases, the formation of N-hydroxylamines can cause DNA damage, and if the dyes are completely reduced to aromatic amines, they can then be oxidized to N-hydroxyderivates by P450 enzymes. In addition, N-hydroxy radicals can be acetylated by enzymes such as O-acetyltransferase, generating nitrenium electrophilic ions which are able to react with DNA forming adducts (Chung et al., 1992; Arlt et al. 2002; Umbuzeiro et al., 2005a). Research carried out by Zbaida (1989) showed that the hydroxylation of non-reactive

Azo Dyes and Their Metabolites: Does the Discharge of the Azo Dye into Water
Bodies Represent Human and Ecological Risks?

107

azo dyes such as azobenzene, increased their binding to microsomal cytochrome P-450 and consequently their rate of reduction (Zbaida, 1989).

Studies with mono-azo dyes have indicated that low electron densities close to the azo bond favor reduction (Walker & Ryan, 1971; Combes & Haveland-Smith, 1982), and this may occur due to hydrogen bonding of an azo N atom together with a proximal naphthol group, producing a keto-hydrazone configuration. It is possible that such a structure, which is present in many food colors, may generate dyes which are more resistant to hepatic and microbial reduction (Parke, 1968 cited as in Combes & Haveland-Smith, 1982).

Kennelly et al. (1984) showed that Direct Black 38, Direct Brown 95, Direct Blue 6, Congo Red, Trypan Blue and Chicago Sky Blue were easily reduced by the intestinal microflora when orally administered to rats by gavage, however when administered via the hepatic portal vein, only Direct Black 38, Direct Brown 95 and Direct Blue 6 were reduced, all of which are potent liver carcinogens (Robens et al., 1980; Chung et al., 1992).

Sweeney et al. (1994) tested azo dyes for genotoxicity following bacterial reduction of the dye. They found that both reduced amaranth and reduced sunset yellow induced cytotoxicity when incubated with a repair deficient E. coli strain in the absence of hepatic enzymes, indicating DNA damage. On the other hand they failed to mutate the S. typhimurium strains TA98 and TA1OO, but in contrast, strain TA102, which detects oxidative mutagens (De Flora et al., 1984), was mutated by reduced amaranth and reduced sunset yellow (Sweeney et al., 1994).

Reduction can also modify the type of activity observed. The direct mutagenicities of Alizarin Yellow GG and Acid Alizarin Yellow R were eliminated by reduction, but in the presence of the exogenous metabolic system (S9), the resulting products were mutagenic and exhibited frame-shift activity (Brown et al., 1978; Combes & Haveland-Smith, 1982).

The Sudan dyes I, II, III and IV are oil-soluble azo dyes (1-amino-2-naphthol-based azo dyes), widely used in coloring plastics, leather, fabrics, printing inks, waxes and floor polishes (An et al., 2007; Xu et al., 2010). Sudan I is a liver and urinary bladder carcinogen in mammals and is also considered as a possible human mutagen, since it can produce the benzenediazonium ion during metabolism catalyzed by cytochrome P450, which could be the mechanism by which Sudan I is activated leading to a carcinogenic final product (Stiborová et al., 2002, 2005; Xu et al., 2010). An et al. (2007) found a dose-dependent increase in DNA migration in the comet assay, and in the frequency of micronuclei with all the concentrations of Sudan I tested (25–100 µM). These data suggest that Sudan I caused breaks in DNA strands and chromosomes. Sudan II causes mutations in Salmonella Typhimurium TA 1538 in the presence of a rat liver preparation (Garner & Nutman, 1977; Xu et al., 2010). Concern about the safety of Sudan III, which is used in cosmetics, has arisen from its potential metabolic cleavage by skin bacteria producing 4-aminoazobenzene and aniline (Pielesz et al., 2002), and Sudan IV has been shown to require reduction and microsomal activation in order to be mutagenic (Brown et al., 1978). These are important mechanisms, since, with the exception of Sudan II, Xu et al. (2010) showed that the bacteria found in the human colon are frequently able to reduce Sudan dyes (Xu et al., 2010).

Almost all the azo dyes are reduced in vivo, but the reduction of the ingested dose is frequently incomplete, and thus a certain amount of the dye can be excreted in the unchanged or conjugated form. For instance, in the case of orally dosing rats with Sudan III, none of the expected reduction products was excreted, although p-aminophenol could be detected in the urine after i.p. injection (IARC, 1975; Combes & Haveland-Smith, 1982). This may have been due to the formation of hydrazones at one of the azo bridges of this diazo dye, which might make it resistant to intestinal reduction (Combes & Haveland-Smith, 1982).

Stahlmann et al. (2006) reported investigations made to evaluate the sensitizing and allergenic potentials of two metabolites expected to be formed by the metabolic activity of skin bacteria and/or by metabolism in the skin. Two metabolites (4-aminoacetanilide and 2-amino-*p*-cresol) of Disperse Yellow 3, an azo dye widely used in the textile industry, were tested using modified local lymph node assay protocols in NMRI mice. The metabolite 2-amino-*p*-cresol gave a clearly positive response in the sensitisation protocol, showing marked increases in lymph node weight and cell proliferation, accompanied by a relative decrease in T-cells and relative increases in B-cells and 1A⁺ cells. Hence, 2-amino-*p*-cresol can be considered to be a stronger allergen in this model. In contrast, 4-aminoacetanilide only led to an increase in lymph node weight and cellularity at the higher concentration of 30%, with no consistent changes in the phenotypic analysis, indicating that this metabolite alone was a weak sensitizer (Stahlmann et al., 2006).

2.4 Dying processing plants effluents and their treatments

The textile industry accounts for two-thirds of the total dyestuff market (Fang et al., 2004; Elisangela et al., 2009). As mentioned before, part of dye used in the textile dyeing process does not attach to the fibers, remaining in the dye baths and eventually being discharged in the wastewater (Fang et al., 2004). The resulting wastewater is usually treated with activated sludge, and the liquid effluent is released to adjacent surface waters (Umbuzeiro et al., 2005 b).

Many dyes do not degrade easily due to their complex structure and textile dye effluent does not decolorize even if the effluent is treated by the municipal wastewater treatment systems (Shaul et al., 1991; Robinson et al., 2002; Forgacs et al., 2004). A study carried out in 1989 showed that the commercial aminoazobenzene dye, C.I. Disperse Blue 79, was not degraded by a conventionally operated activated sludge process and that 85% of the dye remained in the system. Of this 85%, 3% was retained by the primary sludge, 62% by the activated sludge and 20% was found in the final liquid effluent released into the environment (US EPA, 1989; Umbuzeiro et al., 2005 b). The use of an anaerobic system before the activated sludge treatment can result in cleavage of the azo bonds and the release of the corresponding aromatic amines. However, the colourless aromatic amines produced by these anaerobic microorganisms can be highly toxic and carcinogenic (Hu, 1994; Banat et al., 1996; Robinson et al., 2002).

Ekici et al. (2001) tested the stability of selected azo dye metabolites in both activated sludge and water and concluded that they were relatively stable in the aquatic environment and could not be efficiently degraded in wastewater plant systems. With respect to their mutagenicity, Fracasso et al. (1992) showed that dye factory effluents from primary and secondary biological treatments increased their levels of mutagenic activity as compared to the raw (untreated) effluent. The use of activated carbon filtration was beneficial but did not completely remove the mutagenic activity of the final effluent (Fracasso et al., 1992; Umbuzeiro et al., 2005 b).

Azo dyes are usually designed to resist biodegradation under aerobic conditions, the recalcitrance of these compounds being attributed to the presence of sulfonate groups and azo bonds. On the other hand, the vulnerability of reducing the azo bonds by different

Azo Dyes and Their Metabolites: Does the Discharge of the Azo Dye into Water
Bodies Represent Human and Ecological Risks?

109

mechanisms (e.g. biotreatment in anaerobic conditions) could result in the generation of aromatic amines, which are somewhat toxic and carcinogenic (Öztürk & Abdullah, 2006; Bae & Freeman, 2007; Kusic et al., 2011). It should also be mentioned that azo dyes are associated with various health risks to humans, and therefore colored wastewaters should be efficiently treated prior to discharge into the natural water bodies (Kusic et al., 2011).

Several methods are used to decolorize textile effluents including physicochemical methods such as filtration and coagulation, activated carbon and chemical flocculation (Gogate & Pandit, 2004). These methods involve the formation of a concentrated sludge, which, in turn creates a secondary disposal problem (Maier et al., 2004; Elisangela et al., 2009), since these methods merely transfer the pollution from one phase to another, which still requires secondary treatment (Gogate & Pandit, 2004; Kusic et al., 2011). Recently, new biological processes have been developed for dye degradation and wastewater reuse, including the use of aerobic and anaerobic bacteria and fungi (Elisangela et al., 2009).

The decolorizing of azo dyes using a fungal peroxidase system is another promising method (Hu, 1994). The ligninolytics are the most widely researched fungi for dye degradation (Elisangela et al., 2009) and of these, the white-rot fungi have been shown to be the most efficient organisms for the degrading of various types of dye such as azo, heterocyclic, reactive and polymeric dyes (Novotný et al., 2004). These fungi produce lignin peroxidase, manganese-peroxidase and laccase, which degrade many aromatic compounds due to their nonspecific systems (Forgacs et al., 2004; Revankar & Lele, 2007; Madhavi et al., 2007). However, all these processes for the mineralization of azo dyes need to be carried out in a separate process, since the dye compounds cannot be incorporated into the medium, and this would be impractical due to the great volume of wastewater requiring treatment (Hu, 1994). In addition, the long growth cycle and complexity of the textile effluents, which are extremely variable in their compositions, limit the performance of these fungi. Although the stable operation of continuous fungal bioreactors for the treatment of synthetic dye solutions has been achieved, the application of white-rot fungi for the removal of dyes from textile wastewaters still confronts many problems due to the large volumes produced, the nature of the synthetic dyes and the biomass control (Nigam et al., 2000; Mielgo et al., 2001; Robinson et al., 2001b; Elisangela et al., 2009).

Of the chemical methods under development, advanced oxidation processes (AOPs) seem to be a promising option for the treatment of toxic and non-biodegradable organic compounds in various types of wastewater, including the colored ones (Forgacs et al., 2004; Gogate & Pandit, 2004; Kusic et al., 2011). AOPs have received considerable attention due to their potential to completely oxidize the majority of the organic compounds present in the water. AOPs could serve as oxidative pretreatment method to convert non or low-biodegradable organic pollutants into readily biodegradable contaminants (Mantzavinos & Psillakis, 2004; Kusic et al., 2011). The electron beam (EB) treatment is also included in the class of AOPs, and laboratory investigations, pilot-plant experiments and industrially established technology have shown the efficiency of the EB treatment in destroying textile dyes in aqueous solutions (Han et al. ,2002; Pálfi et al., 2011).

Chlorine has been extensively used as a complementary treatment to remove or reduce the color of industrial effluents containing dyes, and also to disinfect the water in drinking water treatment plants (Sarasa et al., 1998; Oliveira et al., 2010). The discoloration process using sodium hypochlorite (NaOCl) or chlorine gas, is based on the electrophilic attack of the amino group, and subsequent cleavage of the chromophore group (responsible for the dye color) (Slokar & Marechal, 1998). However, the treatment of textile effluents using the conventional activated sludge method followed by a chlorination step, is not usually an effective method to remove azo dyes, and can generate products which are more mutagenic than the original untreated dyes, such as PBTAs (chlorinated 2-phenylbenzotriazoles). It has been reported that conventional chlorination should be used with caution in the treatment of aqueous samples contaminated with azo dyes (Umbuzeiro et al., 2005b; Oliveira et al., 2010).

Another alternative could be the use of photoelectrocatalysis on titanium supported nanocrystalline titanium dioxide thin film electrodes, where active chlorine is produced promoting the rapid degradation of reactive dyes (Carneiro et al., 2004; Osugi et al., 2009). Osugi et al. (2009) investigated the decolorizing of the mutagenic azo dyes Disperse Red 1, Disperse Red 13 and Disperse Orange 1 by chemical chlorination and photoelectrochemical oxidation on Ti/TiO_2 thin-film electrodes using NaCl and Na_2SO_4 media. After 1 h of treatment, 100% decolorizing was achieved with all the methods tested. After 1 h of photoelectrocatalytic oxidation, all the dye solutions showed complete reduction of the mutagenic activity using the strains TA98 of *Salmonella* in the absence or presence of the S9 mix, suggesting that this process could be a good option for the removal of disperse azo dyes from aqueous media. The results involving conventional chlorination showed that this method did not remove the mutagenic response from the dyes, and in fact promoted an increase in mutagenic activity in the presence of metabolic activity for Disperse Red 13 (Osugi et al., 2009).

3. Conclusions

The discharge of azo dyes into water bodies presents human and ecological risks, since both the original dyes and their biotransformation products can show toxic effects, mainly causing DNA damage. Azo dyes are widely used by different industries, and part of the dyes used for coloring purposes is discharged into the environment. The azo dyes constitute an important class of environmental mutagens, and hence the development of non-genotoxic dyes and investment in research to find effective treatments for effluents and drinking water is required, in order to avoid environmental and human exposure to these compounds and prevent the deleterious effects they can have on humans and aquatic organisms.

4. Acknowledgements

This work was supported by the Faculty of Pharmaceutical Science at Ribeirão Preto, University of São Paulo, Brazil, and by FAPESP and CAPES.

5. References

Al-Sabti, K. (2000). Chlorotriazine Reactive Azo Red 120 Textile Dye Induces Micronuclei in Fish. *Ecotoxicology and Environmental Safety*, Vol. 47, No. 2, pp.149-155, 2000

An, Y.; Jiang, L.; Cao, J.; Geng, C.; Zhong, L. (2007). Sudan I induces genotoxic effects and oxidative DNA damage in HepG2 cells. *Mutation Research*, Vol. 627, No. 2, pp. 164-70, ISSN 1383-5718

Anliker, R. (1977). Color chemistry and the environment, *Ecotoxicology and Environmental Safety*, Vol. 1, No. 2, pp. 211-237

Anliker, R. (1979). Ecotoxicology of dyestuffs - a joint effort by industry. *Ecotoxicology and Environmental Safety*, Vol. 3, No. 1, pp. 59-74

Arlt, V.M.; Glatt, H.; Muckel, E.; Pabel, U.; Sorg, B.L.; Schmeiser, H.H.; Phillips, D.H. (2002). Metabolic activation of the environmental contaminant 3 nitrobenzanthrone by human acetyltransferases and sulfotransferase. *Carcinogenesis*, Vol. 23, No. 11, pp.1937–1945

Bae, J.S.; Freeman, H.S. (2007). Aquatic toxicity evaluation of new direct dyes to the Daphnia magna. Dyes and Pigments, Vol. 73, No. 1, pp. 81-85, ISSN 0143-7208

Banat, I.M.; Nigam, P.; Singh , D. & Marchant, R. (1996). Microbial decolorization of textile-dyecontaining effluents: a review. *Bioresource Technology*, Vol. 58, No. 3, pp. 217-227, ISSN 0960-8524

Bonser, G.M.; Clayson, D.B.; Jull, J.W. (1963). The potency of 20-methylcholanthrene relative to other carcinogens on bladder implantation. *British Journal of Cancer*, Vol.17, No. 2, pp.235-241

Brown, J.P.; Roehm, G.W.; Brown, R.J. (1978). Mutagenicity testing of certified food colours and related azo, xanthene and triphenylmethane dyes with the Salmonella/microsome system. *Mutation Research*, Vol. 56, No. 1, pp. 249-271 (Abstract)

Carliell, C.M.; Barclay S.J.; Naidoo N.; Buckley C.A.; Mulholland, D.A.; Senior, E. (1995). Microbial decolourisation of a reactive azo dye under anaerobic conditions. Water SA, Vol. 21, pp 61-69

Carneiro, P.A.; Osugi, M.E.; Sene, J.J.; Anderson, M.A.; Zanoni, M.V.B. (2004). Evaluation of color removal and degradation of a reactive textile azo dye on nanoporous TiO2 thin-film electrodes *Electrochimica Acta*, Vol. 49, No. 22-23, pp. 3807-3820, ISSN 0013-4686

Cerniglia, C.E; Zhuo, Z.; Manning, B.W.; Federle, T.W.; Heflich, R.H. (1986). Mutagenic activation of the benzidine-based dye Direct Black 38 by human intestinal microflora. *Mutation Research*, Vol. 175, No. 1, pp. 11-16, ISSN 0165-7992

Chadwick, R.W.; George, S.E.; Claxton, L.D. (1992). Role of the gastrointestinal mucosa and microflora in the bioactivation of dietary and environmental mutagens or carcinogens. *Drug Metabolism Reviews*, Vol. 24, No. 4, pp. 425-492

Chang, J.S.; Chien Chou, C.; Yu-Chih Lin, Y.C.; Ping-Jei Lin, P.J.; Jin-Yen Ho, J.Y. & Hu, T.L.(2001a) Kinetic characteristics of bacterial azo-dye decolorization by *Pseudomonas luteola. Water Research*, Vol. 35, No. 12, pp. 2841–2850.

Chang, J.S.; Chou, C.& Chen, S.Y. (2001b). Decolorization of azo dyes with immobilized *Pseudomonas luteola. Process Biochemistry*, Vol. 36, No. 8-9, pp. 757-763, ISSN 0032-9592

Chequer, F.M.D.; Angeli, J.P.F.; Ferraz, E.R.A.; Tsuboy, M.S.; Marcarini, J.C.; Mantovani, M.S.; Oliveira, D.P. (2009). The azo dyes Disperse Red 1 and Disperse Orange 1 increase the micronuclei frequencies in human lymphocytes and in HepG2 cells. *Mutation Research*, Vol. 676, pp. 83-86, ISSN 1383-5718

Choudhary, G. (1996). Human health perspectives on environmental exposure to benzidine: a review. Chemosphere, Vol. 32, No. 2, pp. 267-291, ISSN 0045-6535(95)00338-X

Chung, K.T.; Fulk, G.E.; Andrews, A.W. (1978). The mutagenicity of methyl orange and metabolites produced by intestinal anaerobes. *Mutation Research,* Vol. 58, No. 2-3, pp. 375-379

Chung, K.T. (1983). The significance of azo-reduction in the mutagenesis and carcinogenesis of azo dyes. *Mutation Research,* Vol. 114, No. 3, pp.269-281, ISSN 0165-111

Chung, K.T. & Cerniglia, C.E. (1992). Mutagenicity of azo dyes: Structure-activity relationships. *Mutation Research/ Reviews in Genetic Toxicology,* Vol. 277, No. 3, pp. 201–220, ISSN 0165-111

Chung, K.T.; Stevens, S.E; Cerniglia, C.E. (1992) The reduction of azo dyes by the intestinal microflora. *Critical Reviews in Microbiology,* Vol. 18, No. 3, pp. 175-190, ISSN 1040-841X

Chung K. T. & Stevens, S. E. (1993). Degradation of azo dyes by environmental microorganisms and helminths. *Environmental Toxicology and Chemistry,* Vol. 12, No. 11, pp. 2121–2132.

Chung, K.T.; Hughes, T.J.; Claxton, L.D. (2000). Comparison of the mutagenic specifity induced by four nitro-group-containing aromatic amines in *Salmonella typhimurium his* genes. Mutation Research, Vol. 465,No. 1-2, pp. 165-171, ISSN 1383-5718

Collier, S.W.; Storm, J.E.; Bronaugh, R.L. (1993). Reduction of azo dyes during in vitro percutaneous absorption. *Toxicology and Applied Pharmacology,* Vol. 118, No. 1, pp. 73-79, ISSN 0041-008X

Combes, R.D.; Haveland-Smith, R.B. (1982). A review of the genotoxicity of food, drug and cosmetic colours and other azo, triphenylmethane and xanthene dyes. *Mutation Research,* Vol. 98, No. 2, pp.101-248, ISSN 0165-1110

De Flora S.; Camoirano A.; Zanacchi C. (1984). Mutagenicity testing with TA97 and TA102 with 30 DNA damaging compounds, negative with other Salmonella strains. Mutation Research, Vol. 134, No. 2-3, pp. 159-165 ISSN 0165-111

De France, B.F.; Carter, M.H.; Josephy, P.D. (1986). Comparative metabolism and mutagenicity of azo and hydrazone dyes in the Ames test. *Food and Chemical Toxicology,* Vol. 24, No. 2, pp.165-169, ISSN 0278-6915

Denisov, I.G.; Makris, T.M.; Sligar, S.G.; Schlichting, I. (2005). Structure and Chemistry of Cytochrome P450. *Chemical Reviews,* Vol. 105, No. 6, pp. 2253-2277, ISSN 10.1021/cr0307143

Dipple, A.; Michejda, C.J.; Weisburger, E.K. (1985) Metabolism of chemical carcinogens. *Pharmacology & Therapeutics,* Vol.27, pp.265-296, ISSN 0163-7258

Ekici, P.; Leupold, G.; Parlar, H. (2001). Degradability of selected azo dye metabolites in activated sludge systems. *Chemosphere,* Vol. 44, No. 4, pp. 721 – 728, ISSN 0045-6535

Elisangela , F.; Andrea, Z.; Fabio, D.G.; Cristiano, R.M.; Regina, D.L.; Artur, C.P. (2009). Biodegradation of textile azo dyes by a facultative *Staphylococcus arlettae* strain VN-

Azo Dyes and Their Metabolites: Does the Discharge of the Azo Dye into Water
Bodies Represent Human and Ecological Risks?

113

11 using a sequential microaerophilic/aerobic process. *International Biodeterioration & Biodegradation*, Vol. 63, No. 3, pp. 280-288, ISSN 0964-8305

Esancy, J.F.; Freeman, H.S.; Claxton, L.D. (1990). The effect of alkoxy substituents on the mutagenicity of some aminoazobenzene dyes and their reductive-cleavage products. *Mutation Research*, Vol. 238, No. 1, pp. 1–22, ISSN 0165-111

Fang, H., Wenrong, H., Yuezhong, L. (2004). Biodegradation mechanisms and kinetics of azo dye 4BS by a microbial consortium. *Chemosphere*, Vol. 57, No. 4, pp. 293-301, ISSN 0045-6535.

Ferraz, E.R.A.; Umbuzeiro, G.A.; de-Almeida, G.; Caloto-Oliveira, A.; Chequer, F.M.D.; Zanoni, M.V.B.; Dorta, D.J.; Oliveira, D.P. (2010). Differential Toxicity of Disperse Red 1 and Disperse Red 13 in the Ames Test, HepG2 Cytotoxicity Assay, and Daphnia Acute Toxicity Test. *Environmental Toxicology*, pp. 1-9, DOI 10.1002/tox.20576

Field, F.E.; Roberts, G.; Hallowes, R.C.; Palmer, A.K.; Kenneth E. Williams, K.E.; Lloyd, J.B. (1977). Trypan blue: identification and teratogenic and oncogenic activities of its coloured constituents. *Chemico-Biological Interactions*, Vol. 16,No. 1, pp. 69-88

Forgacs, E.; Cserháti, T.; Oros, G. (2004). Removal of synthetic dyes from wastewaters: a review. *Environment International*, Vol. 30, No. 7, pp. 953-971, ISSN 0160-4120

Forss, J.; Welander, U. (2011). Biodegradation of azo and anthraquinone dyes in continuous systems. *International Biodeterioration & Biodegradation*, Vol. 65, No. 1, pp. 227-237, ISSN 0964-8305

Fracasso, M.E.; Leone, R.; Brunello, F.; Monastra, C.; Tezza, F.; Storti, P.V. (1992). Mutagenic activity in wastewater concentrates from dye plants. *Mutation Research*, Vol. 298, No. 2, pp. 91-95, ISSN 0165-1218

Fujita, S. & Peisach, J. (1978). Liver microsomal cytochromes *P*-450 and azoreductase activity *The Journal of Biological Chemistry*, Vol. 253, No. 13, pp. 4512-4513

Garg, A.; Bhat, K.L.; Bock, C.W. (2002). Mutagenicity of aminoazobenzene dyes and related structures: a QSAR/QPAR investigation. *Dyes and Pigments*, Vol. 55, No. 1, pp. 35-52, ISSN 0143-7208

Garner, R.C. & Nutman, C.A. (1977). Testing of some azo dyes and their reduction products for mutagenicity using Salmonella typhimurium TA 1538. *Mutation Research*, Vol. 44, No. 1, pp. 9–19

Gogate, R.; Pandit, B. (2004). A review of imperative technologies for wastewater treatment I: Oxidation technologies at ambient conditions. *Advances in Environmental Research*, Vol. 8, pp. 501-551, ISSN 1093-0191

Golka, K.; Kopps, S.; Myslak, Z.W. (2004). Carcinogenicity of azo colorants: influence of solubility and bioavailability – a Review. *Toxicology Letters*, Vol. 151, No. 1, pp. 203-210, ISSN 0378-4274

Han, B.; Ko, J.; Kim, J.; Kim, Y.; Chung, W.; Makarov, I.E.; Ponomarev, A.V.; Pikaev, A.K. (2002). Combined electron-beam and biological treatment of dyeing complex wastewater. Pilot plant experiments. *Radiation Physics and Chemistry*, Vol. 64, No. 1, pp. 53–59, ISSN 0969-806X

Hartman, C.P.; Fulk, G.E.; Andrews, A.W. (1978). Azo reduction of trypan blue to a known carcinogen by a cell-free extract of a human intestinal anaerobe. *Mutation Research*, Vol. 58, No. 2-3, pp. 125-132

Hashimoto, Y.; Watanabe, H.; Degawa, M. (1977). Mutagenicity of methoxyl derivatives of N-hydroxy-4-amino-azobenzenes and 4-nitroazobenzene. *Gann*, Vol. 68, No. 3, pp. 373-374

Haveland-Smith, R. B.; Combes. R. D. (1980a). Screening of food dyes for genotoxic activity. Food and Cosmetics Toxicology, Vol. 18, No. 3, pp. 215-221

Haveland-Smith, R.B.; Combes, R.D. (1980b). Genotoxicity of the food colours Red 2G and Brown FK in bacterial systems: use of structurally-related dyes and azo-reduction. *Food and Cosmetics Toxicology*, Vol.18, No. 3, pp.223-228

Hu, T.L. (1994). Decolourization of reactive azo dyes by transformation with *Pseudomonas Luteola. Bioresource Technology*, Vol. 49, No. 1, pp. 47-51, ISSN 0960-8524

Hunger, K. (1994). On the toxicology and metabolism of azo dyes. *Chimia*, Vol. 48, pp. 520-522, ISSN 0009-4293

IARC (1975). Working Group Monographs on the Evaluation of Carcinogenic Risk of Chemicals to Man, Vol. 8; Some Aromatic Azo Compounds, Int. Agency for Research on Cancer, Lyon

Jin, X.C.; Liu, G.Q.; Xu, Z.H.; Tao, W.Y. (2007). Decolourisation of a Dye Industry Effluent by Aspergillus fumigatus XC6. *Applied Microbiology and Biotechnology*, Vol. 74, pp. 239-243

Kennelly, J.C.; Hertzog, P.J.; Martin, C.N. (1982). The release of 4,4'-diaminobiphenyls from azodyes in the rat. Carcinogenesis, Vol. 3, No. 8, pp. 947–951

Kennelly, J.C.; Aidan Shaw, A.; Martin, C.N. (1984). Reduction to benzidine is not necessary for the covalent binding of a benzidine azodye to rat liver DNA. *Toxicology*, Vol. 32, No. 4, pp. 315-324, ISSN 0300-483X

Khalid, A.; Arshad, M.; Crowley, D.E. (2009). Biodegradation potential of pure and mixed bacterial cultures for removal of 4-nitroaniline from textile dye wastewater.Water Research , Vol. 43, No. 4, pp.1110-1116, ISSN 0043-1354

Kleinow, K.M.; Melancon, M.J.; Lech, J.J. (1987). Biotransformation and Induction: Implications for Toxicity, Bioaccumulation and Monitoring of Environmental Xenobiotics in Fish. *Environmental Health Perspectives*, Vol. 71, pp. 105-119

Križanec, B., Marechal, A.M.L. (2006).Dioxins and dioxin-like persistent organic pollutants in textiles and chemicals in the textile sector. *Croatica Chemica Acta*, Vol. 79, pp.177-186, ISSN 0011-1643

Kučerová, M.; Polivkowi, Z.; Gregor, V.; Dolanská, M.; Málek, B.; Kliment, V.; Žďárský, E.; Maroušková, A.; Nováková, J. (1987). The possible mutagenic effect of the occupation of TV announcer. *Mutation Research*, Vol. 192, No. 1, pp. 59-63, ISSN 0165-7992

Kusic, H.; Juretic, D.; Koprivanac, N.; Marin, V.; Božić, A.L. (2011). Photooxidation processes for an azo dye in aqueous media: Modeling of degradation kinetic and ecological parameters evaluation. *Journal of Hazardous Materials*, Vol. 185, No. 2-3, pp. 1558–1568, ISSN 0304-3894

Azo Dyes and Their Metabolites: Does the Discharge of the Azo Dye into Water
Bodies Represent Human and Ecological Risks?

115

Lima, R.O.A.; Bazo, A.P.; Salvadori, D.M.F.; Rech, C.M.; Oliveira, D.P.; Umbuzeiro, G.A. (2007). Mutagenic and carcinogenic potential of a textile azo dye processing plant effluent that impacts a drinking water source. *Mutation Research*, Vol. 626, No. 1-2, pp. 53-60, ISSN 1383-5718

Livingstone, D.R. (1998) The fate of organic xenobiotics in aquatic ecosystems: quantitative and qualitative differences in biotransformation by invertebrates and fish. *Comparative Biochemistry and Physiology Part A: Molecular & Integrative Physiology*, Vol. 120, No. 1, pp. 43-49, ISSN 1095-6433

Lu, K.; Zhang, X.L.; Zhao, Y.L.; Wu, Z.L. (2010). Removal of color from textile dyeing wastewater by foam separation. *Journal of Hazardous Materials*, Vol. 182, No. 1-3, pp. 928-932, ISSN 0304-3894

Madhavi, S.; Revankar, S.; Lele, S. (2007). Synthetic dye decolorization by white rot fungus, Ganoderma sp. WR-1. *Bioresource Technology*, Vol. 98, No. 4, pp. 775-780, ISSN 0960-8524

Maier, J.; Kandelbauer, A.; Erlacher, A.; Cavaco-Paulo, A.; Gübitz, M.G., (2004). A new alkali-thermostable azoreductase from Bacillus sp. strain SF. *Applied and Environmental Microbiology*, Vol. 70, No. 2, pp. 837–844, ISSN 0099-2240

Manning, B.W.; Cerniglia, C.E.; Federle, T.W. (1985). Metabolism of the benzidine-based azo dye Direct Black 38 by human intestinal microbiot. *Applied and Environmental Microbiology*, Vol. 50, No. 1, pp. 10-15, ISSN 0099-2240

Mansuy, D. (2007). A brief history of the contribution of metalloporphyrin models to cytochrome P450 chemistry and oxidation catalysis. *Comptes Rendus Chimie*, Vol.10, No. 4-5, pp. 392-413, ISSN 1631-0748

Mantzavinos, D. & Psillakis, E. (2004). Enhancement of biodegradability of industrial wastewaters by chemical oxidation pre-treatment. *Journal of Chemical Technology and Biotechnology*, Vol. 79, pp. 431 – 454

Matsushima, T.; Teichman, B.; Samamura, M.; Sugimura, T. (1978). Mutagenicity of azo-compounds, Improved method for detecting their mutagenicities by the Salmonella mutation test. Mutation Research, Vol. 54, No. 2, pp. 220-221 (abstract)

Meal, P.F.; Cocker, J.; Wilson, H.K.; Gilmour, J.M. (1981). Search for benzidine and its metabolites in urine of workers weighing benzidine-derived dyes. *British Journal of Industrial Medicine*, Vol. 38, No. 2, pp. 191-193

Mielgo, I., Moreira, M.T., Feijoo, G., Lema, J.M., 2001. A packed-bed fungal bioreactor for continuous decolourisation of azo-dyes (Orange II). *Journal of Biotechnology*, Vol. 89, No. 2-3, pp. 99-106, ISSN 0168-1656

Miller, J. A., Miller, E. C.; Baumann, C. A. (1945). On the methylation and demethylation of certain carcinogenic azo dyes in the rat. *Cancer Research*, Vol. 5, pp. 162-168

Miller, E. C., Plescia, A. M., Miller, J. A. & Heidelberger, C. (1952). The metabolism of methylated aminoazo dyes. I. The demethylation of 3'-methyl-4-dimethyl-C14-aminoazobenzene in vivo. *The Journal of Biological Chemistry*, Vol. 196, pp. 863-874

Modi, H.A.; Garima Rajput, G.; Ambasana, C. (2010). Decolorization of water soluble azo dyes by bacterial cultures, isolated from dye house effluent. *Bioresource Technology*, Vol. 101, No. 16, pp. 6580–6583, ISSN 0960-8524

Mueller, G. C. & Miller, J. A. (1953). The metabolism of methylated aminoazo dyes. II. Oxidative demethylation by rat liver homogenates. *The Journal of Biological Chemistry*, Vol. 202, pp. 579-587

Nam, S. & Renganathan, V. (2000). Non-enzymatic reduction of azo dyes by NADH. *Chemosphere*, Vol. 40, No. 4, pp. 351-357, ISSN 0045-6535

Nestmann, E.R.; Kowbel, D.J.; Wheat , J.A. (1981) Mutagenicity in Salmonella of dyes used by defence personnel for the detection of liquid chemical warfare agents, Carcinogenesis, Vol. 2, No. 9, pp. 879-883.

Nigam, P.; Armour, G.; Banat, I.M.; Singh, D.; Marchant, R. (2000). Physical removal of textile dyes from effluents and solid-state fermentation of dye-adsorbed agricultural residues. *Bioresource Technology*, Vol. 72, No. 3, pp. 219-226, ISSN 0960-8524

Novotný, C.; Svobodová, K.; Kasinath, A.; Erbanová, P., 2004. Biodegradation of synthetic dyes by *Irpex lacteus* under various growth conditions. *International Biodeterioration & Biodegradation*, Vol. 54, No. 2-3, pp. 215-223, ISSN0964-8305

Oliveira, D.P.; Carneiro, P.A.; Sakagami, M.K.; Zanoni, M.V.B.; Umbuzeiro, G.A. (2007) Chemical characterization of a dye processing plant effluent–Identification of the mutagenic components. *Mutation Research*, Vol. 626, No. 1-2, pp. 135-142, ISSN 1383-5718

Oliveira, G.A.R.; Ferraz, E.R.A.; Chequer, F.M.D.; Grando, M.D.; Angeli, J.P.F.; Tsuboy, M.S.; J.C. Marcarini, J.C.; Mantovani, M.S.; Osugi, M.E.; Lizier, T.M.; Zanoni, M.V.B.; Oliveira, D.P. (2010). Chlorination treatment of aqueous samples reduces, but does not eliminate, the mutagenic effect of the azo dyes Disperse Red 1, Disperse Red 13 and Disperse Orange 1. *Mutation Research*, Vol. 703, No.2, pp. 200-208, ISSN 1383-5718

Osugi, M.E.; Rajeshwar, K.; Ferraz, E.R.A.; Oliveira, D.P.; Araújo, A.R.; Zanoni, M.V.B. (2009). Comparison of oxidation efficiency of disperse dyes by chemical and photoelectrocatalytic chlorination and removal of mutagenic activity. *Electrochimica Acta*, Vol. 54, No. 7, pp. 2086-2093, ISSN 0013-4686

Öztürk, A.; Abdullah, M.I. (2006). Toxicological effect of indole and its azo dye derivatives on some microorganisms under aerobic conditions, Science of the Total Environment, Vol. 358, No. 1-3, pp. 137-142, ISSN 0048-9697

Pálfi, T.; Wojnárovits, L.; Takács, E. (2011). Mechanism of azo dye degradation in Advanced Oxidation Processes: Degradation of Sulfanilic Acid Azochromotrop and its parent compounds in aqueous solution by ionizing radiation. *Radiation Physics and Chemistry*, Vol. 80, No. 3, pp. 462-470, ISSN 0969-806X

Pandey, A.; Singh, P.; Iyengar, L. (2007). Review: Bacterial decolorization and degradation of azo dyes. *International Biodeterioration & Biodegradation* Vol. 59, No. 2, pp. 73-84, ISSN 0964-8305

Pielesz, A. (1999). The process of the reduction of azo dyes used in dyeing textiles on the basis of infrared spectroscopy analysis. *Journal of Molecular Structure* Vol. 511-512, No. 23, pp. 337-344, ISSN 0022-2860

Azo Dyes and Their Metabolites: Does the Discharge of the Azo Dye into Water
Bodies Represent Human and Ecological Risks?

117

Pielesz, A.; Baranowska, I.; Rybak, A.; Włochowicz, A. (2002). Detection and Determination of Aromatic Amines as Products of Reductive Splitting from Selected Azo Dyes. *Ecotoxicology and Environmental Safety*, Vol. 53, No. 1, pp. 42-47, ISSN 0147-6513

Pinheiro, H.M.; Touraud, E.; Thomas, O. (2004). Aromatic amines from azo dye reduction: status review with emphasis on direct UV spectrophotometric detection in textile industry wastewaters. *Dyes and Pigments*, Vol. 61, No.2, pp. 121–139, ISSN 0143-7208

Platzek, T.; Lang, C.; Grohmann, G.; Gi, U.S.; Baltes, W. (1999). Formation of a carcinogenic aromatic amine from an azo dye by human skin bacteria in vitro. *Human & Experimental Toxicology*, Vol. 18, No. 9, pp. 552-559

Prival, M.J. & Mitchell, V.D. (1982). Analysis of a method for testing azo dyes for mutagenic activity in *Salmonella typhimurium* in the presence of flavin mononucleotide and hamster liver S9. *Mutation Research*, Vol. 97, No. 2, pp.103-116, ISSN 0165-116

Rafii, F.; Hall, J. D.; Cerniglia, C.E. (1997). Mutagenicity of azo dyes used in foods, drugs and cosmetics before and after reduction by *Clostridium* species from the human intestinal tract. *Food and Chemical Toxicology*, Vol. 35, No. 9, pp. 897-901, ISSN 0278-6915

Rai, H.; Bhattacharya, M.; Singh, J.; Bansal, T.K.; Vats, P.; Banerjee, U.C. (2005). Removal of Dyes from the Effluent of Textile and Dyestuff Manufacturing Industry: A Review of Emerging Techniques with Reference to Biological Treatment. *Critical Review in Environmental Science and Technology*, Vol. 35, pp. 219-238, ISSN 1064-3389

Rajaguru, P.; Fairbairn, L.J.; Ashby, J.; Willington M.A.; Turner,S.; Woolford, L.A.; Chinnasamy, N.; Rafferty, J.A. (1999). Genotoxicity studies on the azo dye Direct Red 2 using the in vivo mouse bone marrow micronucleus test. *Mutation Research*, Vol. 444, pp.175-180, ISSN 1383-5718

Reid, T. M.; Morton, K. C.; Wang, C. Y.; King C. M. (1983) Conversion of Congo red and 2-azoxyfluorene to mutagens following *in vitro* reduction by whole-cell rat cecal bacteria. *Mutation Research*, Vol. 117, pp. 105-112, ISSN 0165-121

Revankar, M.S.; Lele, S.S. (2007). Synthetic dye decolorization by white rot fungus, *Ganoderma* sp. WR-1. *Bioresource Technology*, Vol. 98, No. 4, pp. 775-780, ISSN 0960-8524

Rinde, E.; Troll, W. (1975). Metabolic reduction of benzidine azo dyes to benzidine in the rhesus monkey. *Journal of the National Cancer Institute*, Vol. 55, No. 1, pp. 181-182

Robens, J. F.; Dill, G. S., Ward, J. M.; Joiner, J. R.; Griesemer, R. A.; Douglas, J. F. (1980). Thirteen-week subchronic toxicity studies of direct blue 6, direct black 38 and direct brown 95 dyes. *Toxicology and Applied Pharmacology*, Vol. 54, pp. 431-442, ISSN 0041-008X

Robinson, T.; McMullan, G.; Marchant, R.; Nigam, P. (2001a). Remediation of Dyes in Textile Effluent: A Critical Review on Current Treatment Technologies with a Proposed Alternative. *Bioresource Technology* Vol. 77, No. 3, pp. 247-255, ISSN 0960-8524

Robinson, T.; Chandran, B.; Nigam, P. (2001b). Studies on the production of enzymes by white-rot fungi for the decolorisation of textile dyes. *Enzyme and Microbial Technology*, Vol. 29, pp. 575–579, ISSN 0141-0229

Robinson, T.; Chandran, B.; Nigam, P. (2002). Removal of dyes from a synthetic textile dye effluent by biosorption on apple pomace and wheat straw. Water Research Vol. 36, No. 11, pp. 2824-2830, ISSN 0043-1354

Saranaik, S., Kanekar, P. (1995). Bioremediation of color of methyl violet and phenol from a dye industry waste effluent using Pseudomonas sp. isolated from factory soil. *The Journal of Applied Bacteriology*, Vol. 79, pp. 459-469

Sarasa, J.; Roche, M.P.; Ormad, M.P.; Gimeno, E.; Puig, A.; Ovelleiro, J.L. (1998). Treatment of a wastewater resulting from dyes manufacturing with ozone and chemical coagulation, Water Research, Vol. 32, No. 9, pp. 2721 - 2727, ISSN 0043-1354

Saratale, R.G.; Saratale, G.D.; Chang, J.S. & Govindwar, S.P. (2011). Bacterial decolorization and degradation of azo dyes: A review. *Journal of the Taiwan Institute of Chemical Engineers*, Vol. 42, No. 1, pp. 138–157, ISSN 1876-1070

Saupe, A. (1999). High-rate biodegradation of 3- And 4-Nitroaniline. *Chemosphere*, Vol. 39, No. 13, pp. 2325-2346, ISSN 0045-653

Shaul, G.M.; Holdsworth, T.J.; Dempsey, C.R.; Dostal, K.A. (1991). Fate of water soluble azo dyes in the activated sludge process. *Chemosphere*, Vol. 22, No.1-2, pp. 107-119, ISSN 1045-535

Shore, J. (1996). Advances in direct dyes. *Indian Journal of Fibers and Textile Research*, Vol. 21, pp. 1-29

Slokar, Y.M.; Marechal, A.M.L. (1998). Methods of decoloration of textile wastewaters. *Dyes and Pigments*, Vol. 37, No. 4, pp. 335-356, ISSN 0143-72081

Sole, G.M. & Chipman, J.K. (1986) The mutagenic potency of chrysoidines and Bismark brown dyes. *Carcinogenesis*, Vol. 7, No. 11, pp. 1921-1923.

Spadaro, J.T.; Gold, M.H.; Renganathan, V., 1992. Degradation of azo dyes by the lignin degrading fungus Phanerochaete chrysosporium. *Applied and Environmental Microbiology*, Vol. 58, No. 8, pp. 2397–2401, ISSN 0099-2240

Stahlmann, R.; Wegner, M.; Riecke, K.; Kruse, M.; Platzek, T. (2006). Sensitising potential of four textile dyes and some of their metabolites in a modified local lymph node assay. *Toxicology*, Vol. 219, No. 1-3, pp. 113–123, ISSN 0300-483X

Stiborová, M.; Martínek, V.; Rýdlová, H.; Hodek, P., Frei, E. (2002). Sudan I is a potential carcinogen for humans: evidence for its metabolic activation and detoxication by human recombinant cytochrome P450 1A1 and liver microsomes. *Cancer Research*, Vol. 62, pp. 5678-5684

Stiborová, M.; Martínek, V.; Rýdlová H.; Koblas, T.; Hodek, P. (2005). Expression of cytochrome P450 1A1 and its contribution to oxidation of a potential human carcinogen 1-phenylazo-2-naphthol (Sudan I) in human livers. *Cancer Letters*, Vol. 220, No. 2, pp. 145-154, ISSN 0304-3835

Sweeney, E.A.; Chipman, J.K.; Forsythe, S.J. (1994).Evidence for Direct-acting Oxidative Genotoxicity by Reduction Products of Azo Dyes. *Environmental Health Perspectives*, Vol. 102, No. 6, pp. 119-122, 1994.

Azo Dyes and Their Metabolites: Does the Discharge of the Azo Dye into Water
Bodies Represent Human and Ecological Risks?

119

Tsuboy, M.S.; Angeli, J.P.F.; Mantovani, M.S.; Knasmüller, S.; Umbuzeiro, G.A.; Ribeiro, L.R. (2007). Genotoxic, mutagenic and cytotoxic effects of the commercial dye CI Disperse Blue 291 in the human hepatic cell line HepG2. Toxicology in Vitro, Vol. 21, No. 8, pp. 1650-1655, ISSN 0887-2333

Umbuzeiro, G.A.; Freeman, H.; Warren, S.H.; Kummrow, F.; Claxton, L.D. (2005a). Mutagenicity evaluation of the commercial product C.I. Disperse Blue 291 using different protocols of the Salmonella assay. Food and Chemical Toxicology, Vol. 43, No. 1, pp. 49-56, ISSN 0278-6915

Umbuzeiro, G.A.; Freeman, H.S.; Warren, S.H.; Oliveira, D.P.; Terao, Y.; Watanabe, T.; Claxton, L.D. (2005b). The contribution of azo dyes to the mutagenic activity of the Cristais River. Chemosphere, Vol. 60, No. 1, pp. 55 - 64, ISSN 0045-6535

US EPA, 1989. Aerobic and anaerobic treatment of C.I. Disperse blue 79. US Department of Commerce, National Technical Information Service (NTIS) (1989) vols. I and II, EPA/600/2-89/051 (PB 90-111642)

Van der Zee, F.P. & Villaverde, S. (2005) Combined anaerobic–aerobic treatment of azo dyes – A short review of bioreactor studies. Water Research, Vol. 39, No. 8, pp. 1425–1440, ISSN 0043-1354

Venturini, S. & Tamaro, M. (1979). Mutagenicity of anthraquinone and azo dyes in Ames' Salmonella typhimurium test, Mutation Research, Vol. 68, No. 4, pp. 307-312.

Walker, R. (1970). The Metabolism of Azo Compounds: A Review of the Literature. Food and Cosmetics Toxicology, Vol. 8, No. 6, pp. 659-676

Walker, R. & Ryan, A.J. (1971) Some molecular parameters influencing rate of reduction of azo compounds by intestinal microflora. Xenobiotica, Vol. 1, No. 4-5, pp. 483-486

Watabe, T.; Ozawa, N.; Kobayashi, F.; Kuruta, H. (1980). Reduction of sulphonated water-soluble azo dyes by micro-organisms from human faeces. Food and Cosmetics Toxicology, Vol.18, No. 4, pp. 349-352 ISSN 0015.6264

Weisburger, J.H. (1997). A perspective on the history and significance of carcinogenic and mutagenic N-substituted aryl compounds in human health. Mutation Research, Vol. 376, No. 1-2, pp. 261–266, ISSN 0027-5107

Weisburger, J.H. (2002). Comments on the history and importance of aromatic and heterocyclic amines in public health. Mutation Research, Vols. 506–507, pp. 9–20, ISSN 0027-5107

Wolff, A. W.; Oehme, F.W. (1974). Carcinogenic chemicals in food as an environmental issue. Journal of the American Veterinary Medical Association, Vol. 164, pp. 623-629

Xu, H.; Heinze, T.M.; Donald D. Paine, D.D.; Cerniglia, C.E.; Chen, H. (2010). Sudan azo dyes and Para Red degradation by prevalent bacteria of the human gastrointestinal tract. Anaerobe, Vol. 16, No. 2, pp. 114-119, ISSN 1075-9964

Zbaida, S.; Stoddart, A.M.; Levine, W.G. (1989). Studies on the mechanism of reduction of azo dye carcinogens by rat liver microsomal cytochrome P-450. Chemico-Biological Interactions, Vol. 69, No. 1, pp.61-71, ISSN 0009-279

Zhao,X & Hardin, I.R. (2007). HPLC and spectrophotometric analysis of biodegradation of azo dyes by Pleurotus ostreatus. Dyes and Pigments, Vol. 73, No. 3, pp. 322-325, ISSN 0143-7208

Zollinger, H (1987). *Colour Chemistry – Synthesis, Properties of Organic Dyes and Pigments*, p. 92-102, VCH Publishers, New York, USA

Zollinger, H. (1991). *Color chemistry: syntheses, properties and applications of organic dyes and pigments.* 3nd ed., Wiley – VCH, ISBN 978 3906 39023 9, New York, USA

6

Functional Suitability of Soluble Peroxidases from Easily Available Plant Sources in Decolorization of Synthetic Dyes

Farrukh Jamal
Department of Biochemistry,
Ram Manohar Lohia Avadh University, Faizabad, U.P.,
India

1. Introduction

With the growing commercial availability of dyes, their range and scope of application is also expanding. Consequently, a large amount of unused dyes are released in the industrial effluents. Ninety percent of textile dyes entering modern activated sludge sewage treatment plants pass through unchanged. There is an intense environmental concern about the fate of these unbound dyes. These discharged dyes form toxic products and their strong color causes turbidity which even at very low concentrations has a huge impact on the aquatic environment.

These synthetic reactive dyes bond covalently with fabric and contain chromophoric groups like anthraquinone, azo, triarylmethane etc. along with reactive groups viz., vinyl sulphone, chlorotriazine, trichloropyrimidine etc. (Sumathi & Manju, 2000; Keharia & Madamvar, 2003). Disperse dyes and acid dyes have low solubility in water. They are mainly used in the dyeing of polyesters and find minor use in dyeing cellulose acetates and polyamides. Azo dyes constitute the largest group of colorants used in industry. These dyes can be precipitated or adsorbed only in small amounts, while under anaerobic conditions they are cleaved by microorganisms forming potentially carcinogenic aromatic amines (Chung & Cerniglia, 1992).

Several studies on decolorization of textile dyes used in industries have been conducted. Some non-textile dyes commonly present in industrial effluents have been studied. Dye removal from wastewaters with traditional physicochemical processes, such as coagulation, adsorption and oxidation with ozone is expensive, can generate large volumes of sludge and usually requires the addition of environmentally hazardous chemical additives (Chen, 2006). On the other hand, most of the synthetic dyes are xenobiotic compounds which are poorly removed by the use of conventional biological aerobic treatments (Marco et al., 2007a, b). Although, biodegradation appears to be a promising technology, unfortunately the analysis of contaminated soil and water has shown persistence of toxic pollutants even in the presence of microorganisms (Robinson et al., 2001; Keharia and Madamvar, 2003).

Decolorization of dye wastewater is an area where innovative treatment technologies need to be investigated. The focus in recent times has shifted towards enzyme based treatment of colored wastewater/industrial textile effluents. The peroxidase and polyphenol oxidases participate in the degradation of a broad range of substrate even at very low concentration. Further, these peroxidases and polyphenol oxidases have been used for treatment of dyes but large scale exploitation has not been achieved due to their low enzymatic activity in biological materials and high cost of purification (Bhunia et al., 2001; Shaffiqu et al., 2002; Verma & Madamwar, 2002). Bioremediation is a viable tool for restoration of contaminated subsurface environments. It is gaining importance due to its cost effectiveness, environmental friendliness and production of less sludge as compared to chemical and physical decomposition processes. Here too microbial treatment has certain inherent limitations (Husain & Jan, 2000; Duran & Esposito, 2000; Torres et al., 2003).

It has been shown that peroxidases catalyze a variety of oxidation reactions and importantly dyes recalcitrant to peroxidase shows significant decolorization in the presence of Redox mediators (Calcaterra et al., 2008). Redox mediated enzyme catalysis has wide application in degradation of polycyclic aromatic hydrocarbons which includes phenols, biphenyls, pesticides, insecticides etc. (Husain & Husain, 2008; Calcaterra et al., 2008).

2. Peroxidases in dye decolorization with Redox mediators

Enzymatic approach has gained considerable interest in the decolorization/degradation of textile and other industrially important dyes present in wastewater. This strategy is ecofriendly and useful in comparison to conventional chemical, physical and biological treatments, which have inherent serious limitations. Enzymatic treatment is very useful due to the action of enzymes on pollutants even when they are present in very dilute solutions and recalcitrant to the action of various microbes participating in the degradation of dyes. Several enzymes (peroxidases, manganese peroxidases, lignin peroxidases, laccases, microperoxidase-11, polyphenol oxidases and azoreductases) have been evaluated for their potential in decolorization and degradation of dyes. Although, some recalcitrant dyes are not degraded/ decolorized in the presence of such enzymes, the addition of certain Redox mediator enhances the range of substrates and efficiency of degradation of the recalcitrant compounds. However, very few Redox mediators are frequently used which includes 1-hydroxybenzotriazole (HOBT), veratryl alcohol, violuric acid (VA), 2-methoxy-phenothiazone. The enzymes in soluble form cannot be exploited on large scale due to limitations of stability and reusability and consequently, the use of immobilized enzymes has significant advantages over soluble enzymes. In the near future, technology based on the enzymatic treatment of dyes present in the industrial effluents/wastewater will play a vital role. Treatment of wastewater on a large scale will also be possible by using reactors containing immobilized enzymes.

2.1 Dye decolorization with turnip (*Brassica rapa*) proteins

The potential of partially purified turnip (*Brassica rapa*) proteins have been studied by Matto and Husain, (2007) on decolorization of certain direct dyes like Direct Red 23, Direct Red 239, Direct Blue 80 and Direct Yellow 4. The turnip proteins showed enhanced decolorization in the presence of Redox mediators. Redox mediators explored for dye decolorization of these dyes were HOBT, alpha naphthol, vanillin, L-histidine, VA, catechol, quinol, bromophenol, 4-nitrophenol and gallic acid. Six out of 10 investigated compounds showed their potential in enhancing the decolorization of direct dyes. The performance of each was evaluated at different concentrations of mediator and enzyme. The decolorization of all tested direct dyes was maximum in the presence of 0.6 mM Redox mediator at pH 5.5

and 30°C. Complex mixtures of dyes decolorized maximally in the presence of 0.6 mM Redox mediator (HOBT/VA). In order to examine the operational stability of the enzyme preparation, the enzyme was exploited for the decolorization of mixtures of dyes for different times in a stirred batch process. There was no change in decolorization of an individual dye or their mixtures after 60 min of incubation; the enzyme caused more than 80% decolorization of all dyes in the presence of 1-hydroxybenzotriazole/violuric acid. However, there was no desirable increase in dye decolorization of the mixtures on overnight incubation. The treatment of such polluted water in the presence of Redox mediators caused the formation of insoluble precipitate, which could be removed by the process of centrifugation. Such catalyzed oxidative coupling reactions may be important for natural transformation pathways for dyes and indicate their potential use as an efficient means for removal of dyes color from waters and wastewaters.

2.2 Dye decolorization with fenugreek (*Trigonella foenum-graecum*) seed proteins

Peroxidase from fenugreek (*Trigonella foenum-graecum*) seeds is highly effective in the decolorization of textile effluent. The role of six Redox mediators has been investigated for effective decolorization by FSP (Fenugreek Seed Proteins). The maximum decolorization of textile effluent was observed in the presence of 1.0 mM 1-hydroxybenzotrizole, 0.7 mM H_2O_2, and 0.4 U/ml of FSP in the buffer of pH 5.0 at 40°C in 2.5 h. The decolorization of textile effluent in a batch process by peroxidase is 85% in 5 h, whereas the complete decolorization of textile effluent by membrane entrapped FSP was observed within 11 h of its operation. The absorption spectra of treated effluent exhibited a marked diminution in the absorbance at different wavelengths compared to untreated effluent. The removal of colored aromatic compounds from wastewater by peroxidases is well-known. These enzymes are currently being employed for the treatment of aromatic compounds (Husain & Husain, 2008; Husain et al., 2009). Most of the studies on dye decolorization have been done in a defined media or synthetic wastewater where a single dye or their mixtures are usually present. However, industrial effluents are more complex due to the presence of other contaminating substances along with colored compounds; under such conditions, the treatment of these pollutants is a difficult problem (Husain et al., 2010). Textile effluents alone are recalcitrant to the action of FSP, but in the presence of Redox mediators, they decolorized significantly. Many other aromatic pollutants such as aromatic amines, dyes and bisphenol A have already been oxidized by peroxidases in the presence of Redox mediators (Karim & Husain, 2009; Matto & Husain, 2008).

The oxidation of effluent in the presence 1.0 mM HOBT is quite low compared to the Redox mediators used in earlier studies to treat various colored pollutants (Rodriguez-Couto et al., 2005; Matto & Husain, 2008). The high concentrations of Redox mediators may not be appropriate for wastewater treatment because of high cost of mediators or possibility of creating negative impacts on effluent toxicity or in the environment upon their disposal into receiving waters (Kurniawati & Nicell, 2007). The maximum decolorization of textile effluent with FSP was at 0.7 mM concentration of H_2O_2. The concentration of H_2O_2 >1.2 mM acted as an inhibitor of peroxidase activity possibly by causing irreversible oxidation of enzyme ferri-heme group which is essential for its activity (Vazquez-Duarte et al., 2001).

The differences in time course of removal of various dyes might be due to the structural barrier and the electron localization among them (Jauregui et al., 2003). The maximum decolorization (68%) of textile effluent was obtained by 0.4 U/ml of FSP. FSP catalyzed the decolorization of effluent over a wide range of pH with an optimum at pH 5.0. The maximum decolorization of effluent was at a temperature of 40°C. FSP successfully removed more than 85% color of the effluent in 5 h. After 5 h, the increase in the rate of effluent decolorization has not been recorded. Complete removal of color has been studied by using enzyme in a membrane bag. This type of reactor would increase the efficiency of dye decolorization because enzyme can be reused as it did not mix with the rest of the treated solution. Some workers reported that *Pseudomonas putida* CCRC14365 peroxidase was capable of removing phenol in hollow-fiber membrane bioreactor. *P. putida* fully degraded 2,000 mg/L of phenol within 73 h; even at a level of 2,800 mg/L, phenol could be degraded by more than 90% after 95 h of operation (Chung et al., 2004). In another report, a novel tube membrane bioreactor has been used for the treatment of an industrially produced wastewater arising in the manufacture of 3-chloronitrobenzene and nitrobenzene; in 1.7 h, over 99% of each 3-chloronitrobenzene and nitrobenzene from the wastewater was degraded (Livingston, 2004).

For the decolorization and removal of colored compounds from textile effluent, spectral analysis is an important aspect to demonstrate a loss in these compounds after treatment with enzyme. The decrease in absorbance peaks in the UV-visible region is a strong evidence for the decolorization of aromatic pollutants from wastewater. Some earlier spectral analyses of the enzymatically treated aromatic amines and their mixtures have shown a significant decrease in absorbance (Kurniawati & Nicell, 2007). There was 80% decolorization of the chromophoric groups and a significant reduction in the peak associated with the aromatic ring. The UV-visible absorbance spectrum for diluted textile effluent has been taken before and after treatment by FSP in the presence of various Redox mediators. The treatment of effluent by FSP in the presence of Redox mediator, HOBT, produces insoluble aggregates which can successfully be removed by centrifugation. The peroxidase from fenugreek seeds has potential in the remediation of hazardous aromatic pollutant and its use could be extended to the large-scale treatment of textile effluents and other related aromatic compounds by employing more effective and cheaper Redox mediators.

2.3 Dye decolorization with Horse Radish Peroxidase (HRP)

HRP is extracted from horse radish roots, and its performance has been evaluated in soluble and immobilized form by conducting batch experiments in the presence of H_2O_2 (Vasantha et al., 2006). The oxidation of Direct Yellow 12 dye has been tested as a function of HRP at fixed concentration of H_2O_2, and at constant HRP activity (1.8 units/ml). The optimum contact time required for dye removal was 1 h 45 min for vials containing 5 ml of dye solution (10 mg/L), 1.8 units of enzyme 1.5 ml/L of H_2O_2 were added and the reaction mixture (24°C, pH 4) was agitated for 2 h 15 min. After this period, dye removal was negligible up to remaining 1 h 35 min.

The Direct Yellow 12 dye at varying aqueous-phase pH of the reaction mixture between 2 and 10 by keeping the dye concentration at 10 mg/L, enzyme concentration at 1.8 units, H_2O_2 dose at 1.5 ml/L, reaction temperature at 24°C and the contact time (1 h 45 min) constant exhibited 70% of the dye removal at an aqueous-phase pH of 4. The dye removal dropped significantly from pH 5 to 8 and the same trend continued up to an aqueous phase of pH 10. Aqueous phase of pH 4 resulted in higher HRP activity compared to other pH ranges from 3 to 9. Hydrogen peroxide acts as a co-substrate to activate the enzymatic action of peroxide radical. It contributed in the catalytic cycle of peroxidase oxidizing the native enzyme to form an enzymatic intermediate which accepts the aromatic compound to carry out its oxidation to form a free radical form. Experiments were carried out to find out the optimum H_2O_2 concentration required to bring out the conversion of dye by varying the H_2O_2 dose from 1 to 3 ml/L in the reaction mixture by keeping all the other experimental conditions constant (dye concentration 10 mg/L; temperature 24°C; enzyme concentration 1.8 units; reaction time 1 h 45 min). Their findings indicate maximum dye removal at H_2O_2 concentration of 2 ml/L and similar degradation with 2.5 and 3 ml/L; therefore, 2 ml/L was taken as the optimum H_2O_2 dose for dye removal (Vasantha et al., 2006).

The concentration of the substrate present in the aqueous phase significantly influences enzyme-mediated reaction. If the amount of enzyme concentration is kept constant and the substrate concentration is gradually increased the reaction will increase until it reaches maximum. After obtaining the equilibrium state, further addition of the substrate will not change the rate of reaction. Studies carried out at different concentrations of the dye, i.e. 5–40 mg/L, keeping the other parameters constant indicated a dye concentration of 25 mg/L to be the cut-off concentration of the dye for optimum removal at the specified experimental conditions along with retaining the activity of HRP enzyme and to protect the protein from denaturation.

Application of free enzyme in industrial process is not economically viable, while immobilization/ entrapment of enzyme results in repeated application. Two types of polymeric materials—alginate and acryl amide—have been used to study their relative efficiency in dye removal for the entrapment of peroxidase. Normally, enzyme immobilization is expected to provide stabilization effect restricting the protein unfolding process as a result of the introduction of random intra- and intermolecular crosslinks. Zille et al. (2003) have reported less availability of the enzyme for interaction with anionic dyes due to the immobilization in a particular matrix. For the immobilization of HRP enzyme acryl amide gel was more efficient in dye removal compared to alginate matrix. About 78% of dye removal was observed with acryl amide gel-immobilized beads, while with alginate matrix it was only 52%. Gel-immobilized HRP was effective in dye removal compared to free HRP (69%), and alginate-immobilized HRP showed inferior performance.

The effect of aqueous-phase pH on the enzyme-catalyzed degradation with alginate- and acryl amide-immobilized HRP enzyme suggests that after pH 4 there is a decrease in the dye removal capacity for both types of the entrapped matrices. About 78% and 54% of the dye removal was observed at an aqueous phase of pH 4 for acryl amide and alginate-entrapped beads respectively. The relative inferior performance of alginate-immobilized HRP compared to acryl amide may perhaps be due to lesser availability of the peroxidase structure to the dye molecule in the alginate mix compared to acryl amide. The effective performance of acryl amide-entrapped beads may be attributed to the nonionic nature of the beads, which results in minimum modification of the enzyme properties and unaffected nature of the charged substrate as well as product diffusion. The electron-withdrawing nature of the azo linkages obstructs the susceptibility of azo dye molecules to oxidative reactions. Only specialized azo dye-reducing enzymes could degrade azo dyes. In addition, when Direct Yellow 12 dye was reduced using the enzymatic method, the oxidation capacity increased with increasing concentrations of HRP and H_2O_2 at pH 4. Gel/alginate based enzyme immobilization reduced the azo dye. The immobilized enzyme beads could further be used two–three times for the removal of the same dye with lower efficiency. The application of enzyme-based systems in waste treatment is unusual, given that many drawbacks are derived from their use, including low efficiency, high costs and easy deactivation of the enzyme.

2.4 Dye color removal with Bitter Gourd (*Momordica charantia*) Peroxidase

Application of partially purified BGP (Bitter Gourd Peroxidase) in the decolorization of textile and other industrially important dyes has been explored. Simple ammonium sulphate precipitated proteins from bitter gourd were taken for the treatment of a number of dyes present in polluted wastewater (Akhtar et al., 2005a, b). Partially purified preparation of BGP was obtained by adding 20–80% ammonium sulphate and this preparation exhibited a specific activity of 99.0 EU (Enzyme units) of peroxidase/mg protein. Peroxidases from bitter gourd were highly stable against pH, heat, urea, water miscible organic solvents, detergents and proteolysis. The dye solution was stable upon exposure to H_2O_2 or to the enzyme alone. The dye precipitation was due to H_2O_2-dependent enzymatic reaction, possibly involving free-radical formation followed by polymerization and precipitation.

Eight textile reactive dyes were tested with increasing concentration (0.133–0.339 EU of BGP/ml) of reaction volume for 2 h at 37°C. The decolorization of Reactive Red 120, Reactive Blue 4, Reactive Blue 160 and Reactive Blue 171 was continuously enhanced by adding increasing concentration of BGP. Reactive Blue 4 was completely decolorized with 0.399 EU of BGP/ml of reaction mixture in the absence of HOBT. Reactive Orange 4, Reactive Orange 86, Reactive Red 11 and Reactive Yellow 84 were recalcitrant to the BGP action. Four reactive textile dyes were treated by step-wise addition of enzyme, adding 0.133 EU of BGP/ml of reaction volume at each step after every 1 h, to the decolorizing solution. Each enzyme addition exhibited rapid disappearance of color. Reactive Blue 160 was completely decolorized after the third enzyme addition while Reactive Blue 4 was almost decolorized completely after the second enzyme addition. Reactive Blue 171 and Reactive Red 120 were decolorized to 56% and 39%, respectively after the third enzyme addition. However, Reactive Red 11, Reactive Orange 4, Reactive Orange 86 and Reactive Yellow 84 were recalcitrant to decolorization by BGP.

Of the reactive dyes incubated with 0.266 EU of BGP/ml of reaction volume for increasing time period only four dyes decolorized on treatment with BGP for 1h at 37°C. Although more color disappeared on incubation for longer duration, the rate of decolorization was slow. All the four dyes were decolorized with varying percentages (30–90%). The decolorization of dyes with 0.266 EU of BGP/ml of reaction volume on incubation at 37°C for 4 h was 88% for Reactive Blue 4, 71% for Reactive Blue 160, 31% for Reactive Blue 171 and 28% for Reactive Red 120. Rest of the four dyes: Reactive Red 11, Reactive Orange 4, Reactive Orange 86 and Reactive Yellow 84 were fully recalcitrant to decolorization by BGP even after 4 h incubation with similar treatment.

Thirteen different non-textile dyes were treated with 0.167 EU of BGP/ml of reaction volume at 37°C for 1h. Carmine, Methyl Orange, Methylene Blue, Coomassie Brilliant Blue G-250, Rhodamine 6G, Methyl Violet 6B, 1:2 naphthaquinone 4-sulphonic acid and Martius Yellow dyes were recalcitrant to decolorization by the BGP action or were slowly decolorized during the progress of the reaction. Maximum decolorization achieved by partially purified BGP was 53% for Naphthalene Black 12B, 57% for Coomassie Brilliant Blue R 250, 96% for Evans Blue, 51% for Eriochrome Black T and 86% for Celestine Blue. Twenty-one dyes used in this study were treated with BGP in the presence of 1.0 mM HOBT and 0.6 mM H_2O_2 at 37°C. Presence of HOBT drastically enhanced the rate of decolorization of recalcitrant dyes. Reactive Orange 4, Reactive Red 11, Reactive Yellow 84 and Reactive Orange 86 were recalcitrant to decolorization in the absence of HOBT. However, these dyes were decolorized up to 98%, 80%, 70% and 78%, respectively by the action of 0.399 EU/ml of BGP in the presence of 1.0 mM HOBT at 37°C for 2 h. Non-textile dyes e.g., Carmine, Methyl Orange, Coomassie Brilliant Blue G250, Rhodamine 6G, Methylene Blue and Methyl Violet 6B, which were recalcitrant to decolorization by BGP in absence of HOBT, were almost completely decolorized in presence of 1.0 mM HOBT. However, 1:2-naphthaquinone 4-sulphonic acid was recalcitrant to decolorization even in the presence of HOBT.

Decolorization of textile reactive dyes in the absence of HOBT occurred by the formation of precipitate, which settled down and removed by centrifugation. Several earlier investigators have shown that the treatment of phenols and aromatic amines by peroxidases and tyrosinases resulted in the formation of large insoluble aggregates. However, the decolorization of textile and other dyes by BGP in presence of 1.0 mM HOBT appeared without the formation of any precipitate. It suggested that the decolorization of dyes took place via degradation of aromatic ring of the compounds or by cleaving certain functional groups as reported elsewhere (Christian et al., 2003; Husain et al., 2009). HOBT could have a dual role, first as a mediator by increasing the substrate range of dyes for BGP and second enhancing the rate of oxidation.

Complex mixtures of various reactive textile and other industrially important dyes have been studied by mixing three or four different dyes in equal proportions and incubated with 0.266 EU of BGP/ml in the presence of 1.0 mM HOBT and 0.6 mM H_2O_2 for 1h at 37°C. Wave length maxima for each dye mixture were determined and decolorization of mixtures was monitored after the incubation period. All the mixtures decolorized by more than 80%. The decolorization rate of mixtures of dyes was slower than that of pure dye solution. This supports an earlier observation that the biodegradation of various phenols in the form of mixtures was quite slow compared to the independent phenol (Kahru et al., 2000). Bourbonnais and Paice (1990) described for the first time the use of Redox mediators by allowing laccase to oxidize non-phenolic compounds thereby expanding the range of substrates that can be oxidized by this enzyme. The mechanism of action of laccase mediator system has been extensively studied and it is used in the textile industry in the finishing process for indigo stained materials. Several workers have demonstrated that the use of Redox mediator system enhanced the rate of dye decolorization by several folds but these mediators were required in very high concentrations (5.7 mM violuric acid/laccase system, 11.6 mM of HOBT/laccase system) (Soares et al., 2001a,b; Claus et al., 2002). For the first time, it was shown that the decolorization of dyes by BGP was effective at very low concentration of HOBT (1.0 mM). The rate of non-textile dyes decolorization was also enhanced by 2–100 folds (Akhtar et al., 2005a). Several reports indicate the enhancement of laccase activity by free radical mediators; however the enhancement of peroxidase activity by a Redox mediator has been studied for the first time. Peroxidase/ Redox mediator system will prove a sensitive and an inexpensive procedure for the treatment of dyes present in complex mixtures/industrial effluent.

2.5 Dye color removal with tomato (*Lycopersicon esculentum*) peroxidase

Matto and Husain (2009) investigated the role of concanavalin A (Con A)-cellulose-bound tomato peroxidase for the decolorization of direct dyes. Cellulose was used as an inexpensive material for the preparation of bioaffinity support. Con A-cellulose-bound tomato peroxidase exhibited higher efficiency in terms of dye decolorization as compared to soluble enzyme under various experimental conditions. Both Direct Red 23 and Direct Blue 80 dyes were recalcitrant to the action of enzyme without a Redox mediator. Six compounds were investigated for Redox-mediating property. Immobilized peroxidase decolorized both dyes to different extent in the presence of all the used Redox mediators. However, 1-hydroxybenzotriazole emerged as a potential Redox mediator for tomato peroxidase catalyzed decolorization of direct dyes. These dyes were maximally decolorized at pH 6.0 and 40°C by soluble and immobilized peroxidase. The absorption spectra of the untreated and treated dyes exhibited a marked difference in the absorption at various wavelengths. Immobilized tomato peroxidase showed a lower Michaelis constant than the free enzyme for both dyes. Soluble and immobilized tomato peroxidase exhibited significantly higher affinity for Direct Red 23 compared to Direct Blue 80.

2.6 Dye decolorization with Pointed Gourd (*Trichosanthes dioica*) Peroxidase

The effects of *Trichosanthes dioica* peroxidase along with Redox mediators on decolorization of water insoluble disperse dyes; Disperse Red 19 (DR19) and Disperse Black 9 (DB9) have been studied by Jamal et al., (2010). Nine different Redox mediators; bromophenol, 2, 4-dichlorophenol, guaiacol, 1-hydroxybenzotriazole, m-cresol, quinol, syringaldehyde, vanillin and violuric acid were evaluated **[Figure-1]**. Among the chosen mediators, 1-hydroxybenzotriazole was most effective for decolorization with PGP (Pointed Gourd Peroxidase). At a concentration of 0.45U/mL the peroxidase could decolorize Disperse Red 19 to a maximum of 79% with O.2mM 1-hydroxybenzotriazole whereas Disperse Black 9 decolorized upto 60% with 0.5mM 1-hydroxybenzotriazole **[Figure-2]**. The time, pH and temperature at which maximum decolorization were recorded was 60 min, 4 and 42°C. It has been observed that the dye solutions were recalcitrant upon exposure to HOBT, H_2O_2 or to the enzyme alone but the enzyme in the presence of Redox mediators was much effective in performing the decolorization of the dyes, implying dye decolorization was a result of Redox mediated H_2O_2-dependent enzymatic reaction. It has already been reported that Redox mediators have the potential to mediate an oxidation reaction between a substrate and an enzyme.

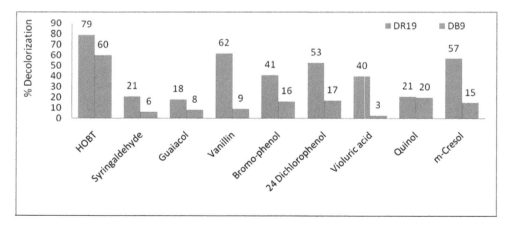

Fig. 1. Percent Dye decolorization as a function of different Redox mediators. The dyes DR19 (25mgL⁻¹,5.0mL) & DB9(50 mgL⁻¹,5.0mL) solutions were incubated independently with PGP (0.45 Uml⁻¹) in the presence of 0.5mM concentration of each Redox mediators; other conditions were 0.8mM H_2O_2, 100mM glycine HCl buffer, pH 4.0 for 60 min at 37°C. (λ_{max} for DR19 and DB9 are 495nm & 461nm respectively) [Jamal et al., 2010]

Different Redox mediators have different mediation efficiency which is governed by Redox potential of the mediator and the oxidation mechanism of the substrate. Oxidation of substrate occurs by free radical formation by the mediator. The free radicals can be formed either by one-electron oxidation of substrate or by abstraction of a proton from the substrate. The pointed gourd peroxidase was effective in decolorizing the dyes at low concentrations of HOBT. Although the extent of decolorization of DR19 and DB9 increased with increasing concentrations of HOBT, the maximum decolorization was observed to be 79% and 60% with 0.2 mM and 0.5 mM for DR19 and DB9, respectively. Further addition of HOBT resulted in a slow decrease in decolorization of both the dyes. This inhibition could likely be due to the high reactivity of HOBT radical, which might undergo chemical reactions with side chains of aromatic amino acid by enzyme thereby; inactivating it. Hence, the dosage of Redox mediator is an important factor for the enzyme-mediated decolorization.

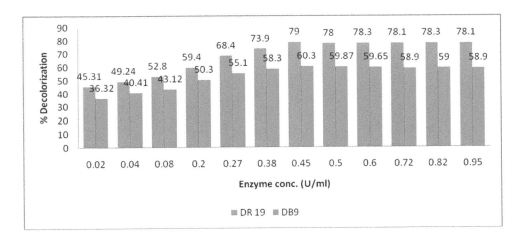

Fig. 2. Percent Dye decolorization as a function of different enzyme (PGP) concentrations. The dyes DR19 (25mgL-1,5.0mL) & DB9(50 mgL-1,5.0mL)solutions were incubated independently with PGP (0.02 to 0.95 Uml-1) in the presence of 0.2mM and 0.5mMconcentration of HOBT for DR19 and DB9 respectively; other conditions were 0.8mM H_2O_2, 100mM glycine HCl buffer, pH 4.0 for 60 min at 37°C. (λ_{max} for DR19 and DB9 are 495nm & 461nm respectively) [Jamal et al., 2010].

The enzyme reacted well to decolorize both the dyes in the presence of 0.8 mM H_2O_2 [Figure-3]. The maximum decolorization was obtained at 0.8mM H_2O_2 which is slightly higher than reported for soybean peroxidase, bitter gourd peroxidase (BGP) and turnip peroxidase. The reaction temperature is an important parameter which effects the decolorization of dyes. The maximum decolorization for both the dyes DR19 and DB9 was at 42°C [Figure-4]. It has been reported that BGP mediated disperse dye decolorization was optimal at 40°C.

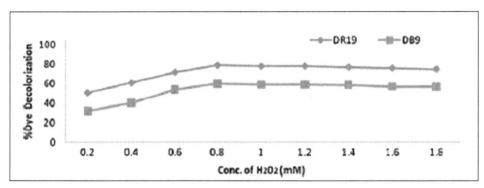

Fig. 3. Percent Dye decolorization at different concentrations of H_2O_2. The dyes DR19 (25mgL-1,5.0mL) & DB9(50 mgL-1,5.0mL) solutions were incubated independently with H_2O_2 concentrations and PGP (0.45 Uml-1) in the presence of varying amounts of H_2O_2 (0.2 to 1.8 mM) for DR19 and DB9 respectively; and other conditions were 100mM glycine HCl buffer, pH 4.0 for 60 min at 37°C. (λ_{max} for DR19 and DB9 are 495nm & 461nm respectively) [Jamal et al., 2010].

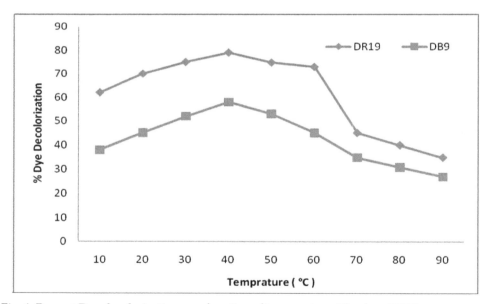

Fig. 4. Percent Dye decolorization as a function of temperature. The dyes DR19 (25mgL-1,5.0mL) & DB9 (50 mgL-1,5.0mL)solutions were incubated independently with PGP (0.45 Uml-1) in the presence of HOBT, 0.8mM H_2O_2, 100mM glycine HCl buffer, pH 4.0 for 60 min at temperatures (20°C to 90°C). (λ_{max} for DR19 and DB9 are 495nm & 461nm respectively) [Jamal et al., 2010].

The maximum decolorization of DR19 and DB9 was obtained at an acidic of pH 4.0 [Figure-5]. It has earlier been reported that the degradation of industrially important dyes by enzymes such as horse radish peroxidase, polyphenol oxidase, BGP and laccase was also maximum in the buffers of acidic pH. DR19 and DB9 were maximally decolorized within 20 min of incubation [Figure-6]. There was slow and gradual enhancement of decolorization upto 60 min of incubation. It was also evident from the observation that DR19 was decolorized to a greater extent within 20 min in the presence of only 0.2 mM HOBT. However, DB9 was decolorized maximally in the presence of 0.5 mM HOBT and decolorization rate was slow. This is consistent with reports that decolorization rate varies, depending upon the type of dye to be treated.

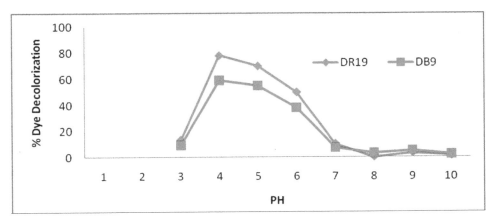

Fig. 5. Percent Dye decolorization as a function of pH. The dyes DR19 (25mgL-1, 5.0mL) & DB9 (50 mgL-1,5.0mL) solutions were incubated independently with PGP (0.45 Uml-1) in the presence of HOBT, 0.8mM H_2O_2, 100mM glycine HCl buffer at different pH (2,3,4,5,6,7,8,9 and 10) for 60 min at 37°C. (λ_{max} for DR19 and DB9 are 495nm & 461nm respectively) [Jamal et al., 2010].

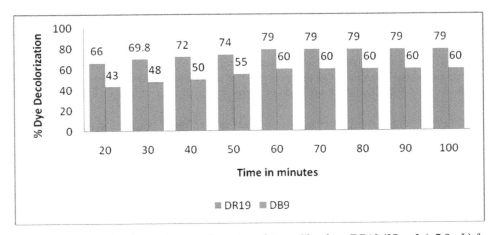

Fig. 6. Percent Dye decolorization as a function of time. The dyes DR19 (25mgL-1, 5.0mL) & DB9(50 mgL-1,5.0mL) solutions were incubated independently with PGP (0.45 Uml-1) in the presence of HOBT, 0.8mM H_2O_2, 100mM glycine HCl buffer, pH 4.0 and for time 20 to 100 min at 37°C. (λ_{max} for DR19 and DB9 are 495nm & 461nm respectively) [Jamal et al., 2010].

2.7 *Trichosanthes dioica* proteins in decolorizing industrially important textile, non-textile dyes and dye mixtures

In one of my recent works, the competency of *Trichosanthes dioica* proteins in decolorizing industrially important textile, non-textile dyes and dye mixtures in the presence of Redox mediators under varying experimental conditions of pH, temperature, time intervals and enzyme concentration on the basis of one-factor-at-a-time (OFAT) method has been studied. All the textile dyes and non-textile dyes, ammonium sulphate, and Tween-20 were procured from Sigma Chemical Co. (St. Louis, MO, USA). Redox mediator's viz., 1-hydroxybenzotriazole (HOBT) and vanillin were obtained from SRL Chemicals (Mumbai, India). All other chemicals were of analytical grade. The pointed gourds were purchased from the local market. The samples were aseptically transferred into sterilized plastic bags.

Briefly, 100 g of pointed gourd was homogenized in 180 ml of 100 mM sodium acetate buffer, pH 5.6. The homogenate was filtered through multi-layers of cheese cloth and then centrifuged at the speed of 10,000 × g on a Remi C-24 cooling centrifuge for 30 min at 4°C. The clear solution thus obtained was used for salt fractionation by adding 10% to 80% (w/v) $(NH_4)_2SO_4$. The proteins were precipitated by continuously stirring at 4°C overnight. The precipitate was collected by centrifugation at 10,000 × g on a Remi C-24 cooling centrifuge, dissolved in 100 mM sodium acetate buffer, pH 5.6 and dialyzed against the assay buffer (0.1 M glycine HCl buffer, pH 4.0) (Akhtar et al., 2005a). Protein concentration was estimated by taking BSA as a standard protein and following the procedure of Lowry et al (1951). Peroxidase activity was determined by a change in the optical density (A_{460} nm) at 37°C by measuring the initial rate of oxidation of 6.0 mM o-dianisidine HCl in the presence of 18.0 mM H_2O_2 in 0.1 M glycine-HCl buffer, pH 4.0, for 20 min at 37°C. One unit of activity was defined as the amount of enzyme that transformed 1μmol of o-dianisidine HCl as substrate per min.

The dyes (45-210 mg/l) were solubilized in 100 mM glycine HCl buffer, pH 4.0. Each dye was independently incubated with pointed gourd peroxidase (PGP) (0.45 EU/mL) in 100 mM glycine HCl buffer, pH 4.0 in the presence of 0.80 mM H_2O_2 for varying times at 37°C. The reaction was stopped by boiling at 100°C for 7 min. Dye decolorization was monitored by measuring the difference at the maximum absorbance for each dye as compared with control experiments without enzyme on UV-visible spectrophotometer (JASCO V-550, Japan). Untreated dye solution (inclusive of all reagents except the enzymes) was used as control (100%) for the calculation of percent decolorization. The dye decolorization was calculated as the ratio of the difference of absorbance of treated and untreated dye to that of treated dye and converted in terms of percentage.

2.7.1 Effect of Redox mediators on decolorization profile of textile and non-textile dyes

The effect of different Redox mediators on the dye decolorization by PGP is shown in **Table-1**. Out of the two different Redox mediators studied for dye decolorization, HOBT was more effective in decolorizing the dyes under study. The extent of decolorization in the presence of HOBT was in the range of 98.6% to 69.8% for the reactive dyes whereas the disperse dyes studied exhibited decolorization in the range of 79.2% to 61.2%. The effective HOBT concentrations were 1.0 mM and 0.2 mM for reactive and disperse dyes, respectively. The dye decolorization with vanillin was 71.2% to 60.2% for reactive dyes and 55.3% to 34.5% for disperse dyes at 1.0 mM concentration. **Table-1** shows the effect of increasing concentrations of HOBT and vanillin (0.05 to 1.5 mM) on textile and other dyes. With increasing concentration of HOBT or vanillin there was an increase in the extent of decolorization of both dyes. However, the decolorization of each textile dye was lower at each of the varying concentration of vanillin than HOBT. The non-textile dyes were effectively decolorized at 1.0 mM of HOBT and vanillin independently but here too, influence of HOBT was substantial. There was not much effect in percent decolorization of the dyes above these concentrations of HOBT or vanillin.

Textile Dyes (λmax)	Percent Dye Decolorization in the presence of *T.diocia* peroxidase and Redox Mediators at varying concentrations (mM) incubated for 2hrs									
	HOBT					Vanillin				
	0.05	0.10	0.20	1.0	1.5	0.05	0.10	0.20	1.0	1.5
Reactive Blue 15 (675 nm)	86.5	89.7	97.6	98.6	98.5	46.6	56.7	64.8	71.2	71.1
Reactive Orange 15 (494 nm)	61.3	67.5	71.2	74.6	74.6	35.4	45.2	52.3	60.7	60.5
Reactive Red 4 (517 nm)	61.6	65.4	67.9	68.2	68.1	32.2	43.2	53.1	61.2	61.1
Reactive Yellow 2 (404 nm)	63.1	65.6	67.6	69.8	69.8	37.6	42.3	54.3	67.2	67.1
Disperse Black 9 (464 nm)	54.2	58.7	69.1	68.7	69.1	15.2	24.2	34.5	55.3	55.2
Disperse Orange 25 (457 nm)	56.1	56.6	61.2	61.1	60.9	28.9	37.8	43.7	45.7	45.6
Disperse Red 19 (495 nm)	67.1	69.9	79.2	78.9	79.1	43.6	56.7	61.2	67.9	67.8
Disperse Yellow 7 (385 nm)	58.9	61.3	67.3	67.2	67.3	15.6	24.8	37.9	37.9	37.8
Non Textile Dyes (λmax)	0.05	0.10	0.20	1.0	1.5	0.05	0.10	0.20	1.0	1.5
Celestine Blue (642 nm)	46.2	52.3	66.8	79.6	79.3	39.7	46.8	61.3	74.9	74.2
Coomassie Brilliant Blue R250 (553 nm)	56.3	69.4	88.9	98.9	98.1	42.1	61.2	76.2	79.8	79.3
Methylene Blue (664 nm)	60.1	67.8	89.3	98.6	98.5	56.7	68.7	71.2	73.2	73.2
Eriochrome Black T (503 nm)	59.3	68.7	90.2	92.5	92.2	45.3	54.5	68.9	79.8	79.7
Evans Blue (611 nm)	67.5	76.4	87.7	97.1	96.8	43.8	61.3	69.9	78.3	78.2
Martius Yellow (430 nm)	6.4	15.6	35.3	39.1	39.1	5.1	13.4	20.9	33.8	33.7
Methyl Orange (505 nm)	64.2	75.6	87.9	98.4	98.4	55.6	67.3	66.7	78.5	78.4
Naphthol Blue Black (618 nm)	57.6	69.7	78.9	97.5	97.2	51.2	62.9	72.1	84.9	84.3
Rhodamine 6G (524 nm)	62.3	69.8	74.3	98.9	98.6	54.3	58.9	68.9	89.4	89.1

Table1. Percent Dye decolorization of different textile and non-textile dyes in the presence *T. dioica* peroxidase along with two different Redox mediators [HOBT and Vanillin].

2.7.2 Enzyme activity profile of *T. dioica* peroxidase mediated decolorization of textile dyes

Figure-7a shows the extent of decolorization of reactive and disperse dyes with increasing concentration of PGP. The maximal decolorization for these two different classes of dyes was observed at PGP concentration of 0.45 EU/ml after an incubation time of 4 h. Dye decolorization was not significantly exhibited with any further increase of PGP. The presence of 0.2 mM and 1.0 mM HOBT for decolorization of disperse and reactive dyes was effective in disappearance of Reactive Blue 15 and Reactive Orange 15 to the extent of 96.2% and 94.6% respectively whereas the others, Reactive Red 4 and Reactive Yellow 2 disappeared up to 88.2% and 89.8% respectively. Among the disperse dyes only Disperse Black 9 was effectively decolorized up to 79% whereas the others showed disappearance of color below 69.3%.

In the presence of 1.0 mM vanillin, partially purified *T. dioica* peroxidase catalyzed decolorization at 0.45 EU/ml after an incubation period of 4 h in the range of 61.4% to 71.3% for the dyes under study [**Figure-7b**]. The decolorization profile for all the dyes significantly increased after the first increase in peroxidase concentration in a period of 2 h. There was no significant change on dye decolorization either on increasing peroxidase concentration or the incubation time.

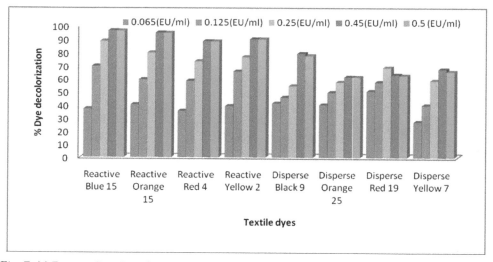

Fig. 7. (a) Percent Dye Decolorization of Textile Dyes in the presence of fixed concentration of HOBT [0.2 mM for Reactive Dyes and 1.0 mM for Disperse Dye] and increasing concentration of *T. dioica* peroxidase enzyme (EU/ml). Please see Table-1 for λ_{max} of each dye.

2.7.3 H_2O_2 and pH activity profile of decolorization of textile dyes

Figure-8 shows that the percent decolorization improved with the increasing concentration of H_2O_2 and the maximum decolorization was observed at a concentration of 0.8 mM and 1.0 mM of H_2O_2 for disperse and reactive dyes respectively which remained substantially unaffected till 1.2 mM H_2O_2.

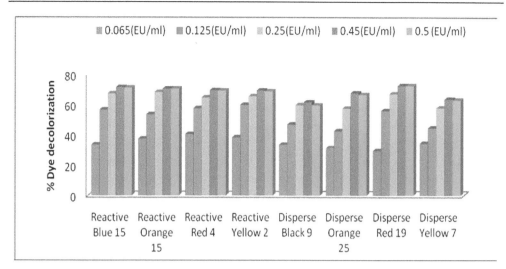

Fig. 7. (b) Percent Dye Decolorization of Textile Dyes in the presence of fixed concentration of vanillin [0.2 mM for Reactive Dyes and 1.0 mM for Disperse Dye] and increasing concentration of *T. dioica* peroxidase enzyme (EU/ml). Please see Table-1 for λ_{max} of each dye.

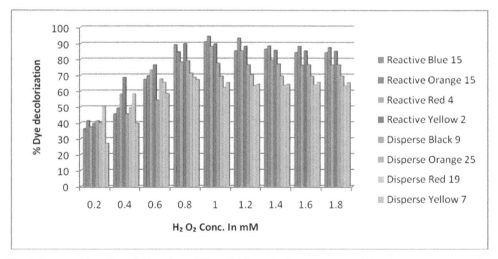

Fig. 8. Percent Dye Decolorization of Textile Dyes in the presence of fixed concentration of HOBT, *T. dioica* peroxidase enzyme (EU/ml) and varying concentration of H_2O_2. Please see Table-1 for λ_{max} of each dye.

To find out the range of pH in which significant decolorization was observed; buffers in the range of pH 2.0 to pH 10.0 were used. The percent decolorization is shown in [**Figure-9**]. An acidic range of pH (3.0 to 6.0) was better suited for dye decolorization. Maximum decolorization was observed at pH 4.0 and pH 5.0 at fixed concentration of *T. dioica* peroxidase and HOBT for disperse and reactive dyes respectively. There was significant decrease in the extent of decolorization in an alkaline medium and at pH 10.0 the decolorization action of the enzyme was almost insignificant / lost.

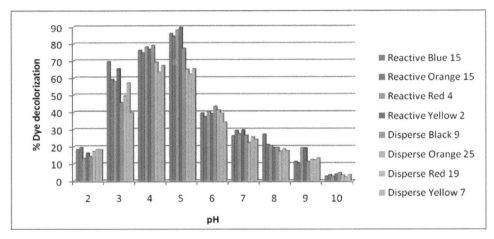

Fig. 9. Percent Dye Decolorization of Textile Dyes in the presence of fixed concentration of HOBT, *T. dioica* peroxidase enzyme (EU/ml) and varying pH. Please see Table-1 for λ_{max} of each dye.

2.7.4 Temperature activity profile of decolorization of textile dyes
The percent decolorization was plotted as a function of temperature and the results are shown in [**Figure-10**]. Among the textile dyes the reactive dyes exhibited maximum decolorization at 50°C whereas the disperse dyes showed maximum decolorization at 40°C in the presence of 1.0 mM and 0.2 mM HOBT. Reactive Blue 15 (96.1%), Reactive Orange 15 (94.4%), Reactive Red 4 (85.2%) whereas disperse dyes under study decolorized in the range of 61.2% to 79%.

2.7.5 Time activity profile of decolorization of textile dyes
The extent of decolorization of textile dyes as a function of time is shown in [**Figure-11**]. Maximum decolorization of reactive and disperse dyes were observed within 1 h of incubation at 50°C and 40°C with 0.45 EU/ml of PGP and 1.0 mM and 0.2 mM HOBT respectively. However, further decolorization of these dyes progressed slowly upto 4 h, although no effective increase was observed even when the dyes were further incubated for longer times. Among the reactive dyes Reactive Blue 15 decolorized almost completely and Reactive Orange 15 up to 78.3% at 4 h incubation, whereas Reactive Red 4 and Reactive Yellow 2 showed decolorization up to 86.3% and 81.2% under similar conditions. The disperse dyes were comparatively resistant to decolorization and only in the presence of HOBT Disperse Red 19, Disperse Yellow 7, Disperse Black 9 decolorized to 79.5%, 72.3% and 71.6% respectively. Disperse Orange 15 was comparatively degraded and decolorized to a lesser extent under similar conditions with a maximum of 61.2%.

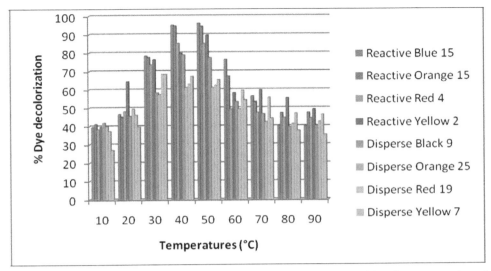

Fig. 10. Percent Dye Decolorization of Textile Dyes in the presence of fixed concentration of HOBT, *T. dioica* peroxidase enzyme (EU/ml) at different temperatures. Please see Table-1 for λ_{max} of each dye.

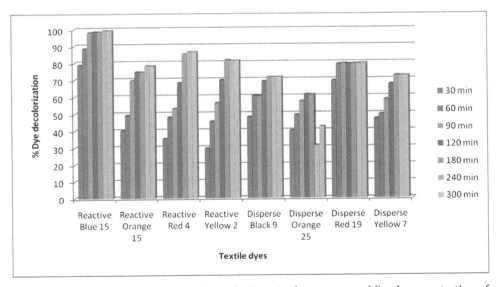

Fig. 11. Percent Dye Decolorization of Textile Dyes in the presence of fixed concentration of HOBT, *T. dioica* peroxidase enzyme (EU/ml) at different time interval. Please see Table-1 for λ_{max} of each dye.

2.7.6 Decolorization profile of non-textile dyes by *T. dioica* peroxidase as a function of enzyme concentration and time

Nine different non-textile dyes were studied. These dyes were treated with different amount of *T. dioica* peroxidase in the range of 0.065 EU/ml to 0.50 EU/ml and incubated for varying time interval with and without the Redox mediator HOBT (1.0 mM) at 37°C as shown in [Table-2]. The results indicated that few non-textile dyes viz., Methylene Blue, Martius Yellow, Methyl Orange, Rhodamine 6G were highly recalcitrant to decolorization by *T. dioica* in the absence of HOBT after 3h of incubation. However, the decolorization progressed slowly with the addition of 1.0 mM HOBT and percent decolorization achieved for Methylene Blue, Martius Yellow and Methyl Orange, was 98.6%, 35.1% and 98.7% respectively; whereas Rhodamine 6G almost decolorized completely. For other dyes, the maximum decolorization exhibited after 3h was 98.7%, 97.3%, 97.1%, 67.8% for Coomassie Brilliant Blue R 250, Naphthol Blue Black, Evans Blue and Eriochrome Black T respectively.

Non Textile Dyes (λmax) nm	0.065 EU/ml		0.125 EU/ml		0.250 EU/ml		0.45 EU/ml		0.50 EU/ml	
	-HOBT (60min)	+ HOBT (60min)	-HOBT (120min)	+HOBT (120min)	-HOBT (160min)	+HOBT (160min)	-HOBT (180min)	+HOBT (180min)	-HOBT (240min)	+HOBT (240min)
Celestine Blue (642 nm)	46.6	57.3	49.7	68.7	54.3	69.3	55.6	79.5	58.9	79.6
Coomassie Brilliant Blue R250 (553 nm)	47.6	92.4	48.2	93.2	51.3	98.9	51.2	98.7	51.1	98.7
Methylene Blue (664 nm)	2.5	67.2	3.7	78.5	3.6	98.4	3.5	98.6	3.2	98.9
Eriochrome Black T (503 nm)	48.7	67.7	51.6	77.4	55.4	91.3	55.3	92.0	55.1	92.2
Evans Blue (607nm)	86.4	94.3	87.5	95.4	87.8	96.7	87.2	97.1	86.7	97.2
Martius Yellow (433 nm)	10.2	33.2	12.4	34.5	13.2	35.6	13.5	35.1	12.3	35.2
Methyl Orange (464 nm)	12.3	87.2	13.2	89.2	13.2	98.5	13.1	98.7	12.2	98.6
Naphthol Blue Black (618 nm)	47.2	78.8	49.2	86.2	56.6	97.4	57.2	97.3	57.9	97.1
Rhodamine 6G (525 nm)	11.2	78.4	13.2	86.3	13.2	99.7	14.2	99.6	13.2	99.5

Table 2. Decolorization of Non textile dyes by *T. dioica* peroxidase at different enzyme concentration, incubation period and with fixed concentration of HOBT, H_2O_2.

2.7.7 Decolorization profile of different dye mixtures by *T. dioica* peroxidase

To simulate the decolorization of dyes from industrial effluent, complex mixtures of textile dyes including reactive, disperse and non-textile dyes were prepared by mixing four different dyes in equal proportions and incubated with 0.45 EU/ml of *T. dioica* peroxidase in the presence of 1.0 mM HOBT and 1.0 mM H_2O_2 for 2 h at 40°C. The decolorization was recorded at wave length maxima of each mixture determined spectrophotometrically. The combinations of different dyes showed decolorization by more than 82% [Figure-12]. The rate of decolorization of dye mixture was slower in comparison to that of individual dyes both in the presence and absence of HOBT. However, the HOBT mediated dye decolorization was more effective.

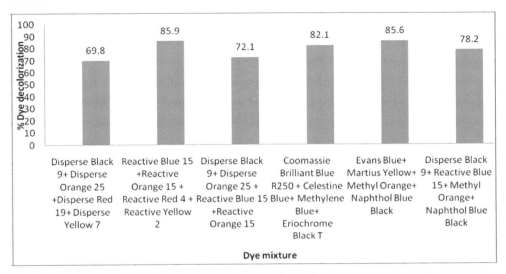

Fig. 12. Degradation/ decolorization of dye mixture by *T. dioica* in the presence of HOBT/H2O2 to mimic dyes in industrial effluents. [Disperse Black 9+ Disperse Orange 25 +Disperse Red 19+ Disperse Yellow 7 (λ_{461}); Reactive Blue 15 +Reactive Orange 15 + Reactive Red 4 + Reactive Yellow 2(λ_{531}); Disperse Black 9+ Disperse Orange 25 + Reactive Blue 15 +Reactive Orange 15(λ_{521}); Coomassie Brilliant Blue R250 + Celestine Blue+ Methylene Blue+ Eriochrome Black T(λ_{595}); Evans Blue+ Martius Yellow+ Methyl Orange+ Naphthol Blue Black (λ_{541}); Disperse Black 9+ Reactive Blue 15+ Methyl Orange+ Naphthol Blue Black (λ_{571})].

We have earlier reported the potential of a novel *T. dioica* plant proteins in the decolorization of disperse dyes (Jamal et al., 2010). In this paper we intended to widen the spectrum of industrially important textile dyes and included some well known non-textile dyes to study the extent of decolorization in the presence of Redox mediators. The cost of an enzyme depends on its degree of purity hence we opted to use ammonium sulphate precipitated proteins from *T. dioica* to study dye decolorization/degradation. The PGP was partially purified using 10% to 80% ammonium sulphate which retained a specific activity of 96 U/mg of protein. The experiments were performed at different enzyme concentration, pH, temperature, incubation period and Redox mediators namely, HOBT and vanillin. An interesting data profile was obtained for the assessment of this enzyme suitability to treat wastewater contaminated with these dyes.

The reactive dyes underwent decolorization by the formation of precipitate which disappeared in the presence of 1.0 mM HOBT. This finding supports earlier reports that treatment of phenols and aromatic amines by peroxidases resulted in formation of large insoluble aggregates (Wada et al., 1995; Tatsumi et al., 1996; Husain & Jan, 2000; Duran & Esposito, 2000). The other dyes studied showed no formation of precipitate during decolorization in the presence of HOBT. This observation supports earlier view that decolorization of dyes took place via degradation of aromatic ring of the compounds or by cleaving certain functional groups (Akhtar et al, 2005a,b; Satar and Husain, 2009). All the other textile dyes showed insignificant or no decolorization with *T. dioica* peroxidase alone when studied under optimum conditions of dye decolorization in the absence of HOBT (data not shown).

The decolorization profile for all the dyes increased significantly at HOBT concentration of 0.2 mM [Table-1]. Upon increasing the HOBT concentration further to 1.5 mM decolorization increased marginally. The reactive dyes under study exhibited decolorization maximally at 1.0 mM HOBT whereas disperse dyes showed maximum decolorization at lower values of HOBT suggesting that reactive dyes are comparatively more resistant to degradation. Vanillin was able to decolorize both the reactive and disperse dyes at 1.0 mM concentration but the extent of decolorization was sufficiently poor in comparison to HOBT for certain dyes like Reactive blue 15, Reactive orange 15 and most of the disperse dyes studied. Disperse Yellow 7 decolorized poorly (37.9%) with *T. dioica* peroxidase in the presence of relatively higher concentration of vanillin.

The non textile dyes exhibited remarkable decolorization in presence of both the Redox mediators at 1.0 mM concentration. There was no significant change on dye decolorization of non textile dyes at concentrations above 1.0 mM. It has already been reported that Redox mediators have the potential to mediate oxidation reaction between a substrate and an enzyme (d'Acunzo et al., 2006). Different Redox mediators have different mediation efficiency which is governed by Redox potential of the mediator and the oxidation mechanism of the substrate (Baiocco et al., 2003). Oxidation of substrate occurs by free radical formation by the mediator. The free radicals can be formed either by one-electron oxidation of substrate or by abstraction of a proton from the substrate (Xu et al., 2001; Fabbrini et al., 2002). In this study, Redox-mediating property of two different compounds as peroxidase mediators was evaluated and extensive study on textile, non-textile and dye mixtures has been performed with HOBT.

Laccase have been used with Redox mediators to oxidize non-phenolic compounds (Bourbonnais & Paice, 1990). The mechanism of action of laccase mediator system has been extensively studied and it is used in the textile industry in the finishing process for indigo stained materials. Several workers have demonstrated that the use of Redox mediator system enhanced the rate of dye decolorization by several folds but these mediators were required in very high concentrations (Soares et al., 2001a,b; Claus et al., 2002). In the present study we have shown the decolorization of both textile and non-textile dyes as well as dye mixtures mediated by *T. dioica* peroxidase under the influence of low concentration of Redox mediators. The peroxidase reacted well to decolorize at concentration of 0.8 mM and 1.0 mM of H_2O_2 for disperse and reactive dyes respectively and remained substantially unaffected till 1.2 mM [Figure-8]. Although the concentration of H_2O_2 greater than 0.75 mM acted as an inhibitor of peroxidase activity by irreversibly oxidizing the enzyme ferri-heme group essential for peroxidase activity (Satar and Husain, 2009), in this study we observed a relatively higher working concentration of H_2O_2 . The enzyme in the presence of Redox mediators was much effective in performing the decolorization of the dyes, implying dye decolorization was a result of Redox mediated H_2O_2-dependent enzymatic reaction. Our results are consistent and very near to values reported earlier for maximum functional concentration of H_2O_2 (Camarero et al., 2005).The enzyme could function better in an acidic medium of pH range 3-6, whereas its decolorizing/degrading activity was adversely affected in alkaline medium [Figure-9]. This finding supports the earlier view of disperse dye decolorization in acidic medium by *T. dioica* and reports further that acid medium favours catalysis of reactive dyes too by this peroxidase. It has earlier been reported that the degradation of industrially important dyes by enzymes such as horse radish peroxidase, polyphenol oxidase, BGP and laccase was also maximum in the buffers of acidic pH (Galindo et al., 2000).

The reaction temperature is an important parameter which effects the decolorization of dyes. The maximum decolorization for both the reactive and disperse dyes were in the temperature range of 40°C to 50°C in the presence of fixed concentration of HOBT [Figure-10]. The extent of decolorization was remarkable for Reactive Blue 15, Reactive Orange 15, Reactive Red 4 at 50°C whereas disperse dyes could be decolorized in the range of 61.2% to 79% at 40°C. The decolorization varies with the nature of the dyes but Redox mediated decolorization with *T. dioica* peroxidase was a better solution for effective decolorization of recalcitrant compounds. The rate of decolorization varied with time and maximum decolorization in the presence of HOBT was observed within one hour of incubation for Reactive Blue15 and Disperse Red 19 [Figure-11]. However, for other dyes the extent of decolorization progressed slowly and reached a plateau after 4 h of incubation. This data is consistent with reports that decolorization rate varies, depending upon the type of dye to be treated (Camarero et al., 2005).

The non-textile dyes studied for decolorization with *T. dioica* peroxidase exhibited enhanced decolorization in the presence of 1.0 mM HOBT, whereas in the absence of Redox mediator decolorization was much slower. Coomassie Brilliant Blue R250, Methylene Blue, Eriochrome Black T, Martius Yellow, Methyl Orange, Naphthol Blue Black and Rhodamine were extensively decolorized by *T. dioica* peroxidase under the influence HOBT [Table-2].The decolorization of Evans blue was not significantly affected by the presence of Redox mediator, although in the presence of HOBT decolorization achieved was higher. The performance of this system was maximal during 160 min of incubation. Celestine blue was decolorized to 79.5% in the presence of HOBT at higher concentration of enzyme as well as longer incubation time. The data in [Table-2] is suggestive of *T. dioica* peroxidase in conjunction with low concentration of HOBT to be wonderful decolorization/degradation system for non-textile dyes as well. Further, dye mixtures simulating industrial effluents exhibited more than 82% decolorization with 1-hydroxybenzotriazole [Figure-12]. The rate of decolorization of dye mixture was slower in comparison to that of individual dyes both in the presence and absence of HOBT. The application of inexpensive peroxidase from easily available source can overcome the limitations in current wastewater treatment strategies. The use of peroxidases can be extended to large-scale treatment of wide spectrum of structural dyes by using immobilized PGP along with relatively cheaper Redox mediators. Thus, the study demonstrated that the peroxidase enzyme isolated from *T. dioica* can be coupled with Redox mediator into a system that can serve as an effective biocatalyst for the treatment of effluents containing recalcitrant dyes from textile, dyeing and printing industries.

3. Future perspective

Dye wastewater discharged from textile and dyestuff industries needs to be treated due to their impact on water bodies and to address growing public concern over their toxicity and carcinogenicity. Many different and complicated molecular structures of dyes make dye wastewater treatment difficult by conventional biological and physico-chemical processes. Therefore innovative treatment technologies need to be investigated. The studies performed using peroxidases from different sources indicates that novel enzyme systems can be created to decolorize wide spectrum of textile, non-textile dyes and dye mixtures under varying set of conditions. The efficacy of decolorization drastically improves with Redox mediators and dyes/dye mixtures recalcitrant to peroxidase exhibited remarkable decolorization. The application of inexpensive peroxidases from easily available sources can overcome the limitations in current wastewater treatment strategies. The use of peroxidases can be extended to large-scale treatment of wide spectrum of structural dyes by using immobilized peroxidases along with relatively cheaper Redox mediators.

4. References

Akhtar, S., Khan, A.A. & Husain, Q. (2005a). Partially purified bitter gourd (*Momordica charantia*) peroxidase catalyzed decolorization of textile and other industrially important dyes. *Bioresour. Technol*, 96, 1804– 1811, ISSN: 0960-8524.

Akhtar, S., Khan, A.A. & Husain, Q. (2005b). Potential of immobilized bitter gourd (*Momordica charantia*) peroxidases in the decolorization and removal of textile dyes from polluted wastewater and dyeing effluent. *Chemosphere*, 60, 291–301, ISSN: 0045-6535.

Baiocco, P., Barreca, A.M., Fabbrini, M., Galli, C. & Gentili, P. (2003). Promoting laccase activity towards non-phenolic substrates: a mechanistic investigation with some laccase-mediator systems. *Org. Biomol. Chem* 1, 191–197, ISSN: 1477-0520.

Bhunia, A., Durani, S. & Wangikar, P.P. (2001). Horseradish peroxidase catalyzed degradation of industrially important dyes. *Biotechnol. Bioeng*, 72, 562–567, ISSN: 0006-3592.

Bourbonnais, R. & Paice, M.G. (1990). Oxidation of non-phenolic substrates. An expanded role for laccase in lignin biodegradation. *FEBS Lett*, 267, 99–102, ISSN: 0014-5793.

Calcaterra, A., Galli, C. & Gentili, P. (2008). Phenolic compounds as likely natural mediators of laccase : A mechanistic assessment. *J. Mol. Cat. B: Enzym*, 51,118-120, ISSN: 0039-7911.

Camarero, S., Ibarra, D., Martı́nez, M.J. & Martı́nez, A.J. (2005). Lignin derived compounds as efficient laccase mediators for decolorization of different types of recalcitrant dyes. *Appl. Environ. Microbiol*, 71, 1775–1784, ISSN: 1432-0614.

Chen, H. (2006). Recent advances in azo dye degrading enzyme research. *Curr. Protein Pept. Sci*, 7,101–111, ISSN: 1389-2037.

Christian, V.V., Shrivastava, R., Novotny, C. & Vyas, B.R. (2003). Decolorization of sulfonaphthalein dyes by manganese peroxidase activity of the white-rot fungus *Phanerochaete chrysosporium*. *Folia Microbiol*, 48, 771–774, ISSN: 1874-9356.

Chung, K.-T. & Cerniglia, C.E. (1992). Mutagenicity of azo dyes: structure activity relationships. *Mutat. Res*, 277, 201–220, ISSN: 0027-5107.

Chung, T. P., Wu, P. C. & Juang, R. S. (2004). Process development for degradation of phenol by Pseudomonas putida in hollow-fiber membrane bioreactors. *Biotechnol. Bioeng*, 87, 219–227, ISSN:0006-3592.

Claus, H., Faber, G. & Konig, H. (2002). Redox-mediated decolorization of synthetic dyes by fungal laccases. *Appl Microbiol Biotechnol*, 59, 672–678, ISSN: 0175-7598.

d'Acunzo, F., Galli, C., Gentili, P. & Sergi, F. (2006). Mechanistic and steric issues in the oxidation of phenolic and non-phenolic compounds by laccase or laccase mediator systems. The case of bifunctional substrates. *New J. Chem*, 30, 583-591, ISSN: 1144-0546.

Duran, N. & Esposito, E. (2000). Potential applications of oxidative enzymes and phenoloxidase-like compounds in wastewater and soil treatment: a review. *Appl. Catal. B: Environ*, 28, 83–99, ISSN: 0926-3373.

Fabbrini, M., Galli, C. & Gentili, P. (2002). Comparing the catalytic efficiency of some mediators of laccase. *J. Mol. Catal. B: Enzy*, 16, 231-240, ISSN: 0039-7911.

Galindo, C., Jaques, P. & Kalt, A. (2000). Photodegradation of the aminobenzene acid orange 52 by three AOPs: UV/H2O2, UV/TiO2 and VIS/TiO2. Comparative mechanistic and kinetic investigations. *J. Photochem. Photobiol. A Chem*, 130, 35–47, ISSN: 1010-6030.

Husain, M. & Husain, Q. (2008). Applications of Redox mediators in the treatment of organic pollutants by using oxidoreductive enzymes. *Critical Reviews in Environmental Science and Technology*, 38, 1–42, ISSN: 1064-3389.

Husain, Q. & Jan, U. (2000). Detoxification of phenols and aromatic amines from polluted wastewater by using polyphenol oxidases. *J. Sci. Ind. Res*, 59, 286–293, ISSN: 0022-4456.

Husain, Q., Husain, M. & Kulshrestha, Y. (2009). Remediation and treatment of organopollutants mediated by peroxidases. *Critical Reviews in Biotechnology*, 29, 94–119, ISSN: 0738-8551.

Husain, Q., Karim, Z. & Banday, Z.Z. (2010). Decolorization of textile effluent by soluble fenugreek (*Trigonella foenum graecum* L) seed peroxidase. *Water Air Soil Pollut*, 212, 319-328, ISSN: 1567-7230.

Jamal, F., Pandey, P.K. & Qidwai, T. (2010). Potential of Peroxidase Enzyme from *Trichosanthes dioica* to mediate Disperse Dye Decolorization in conjunction with Redox Mediators. *J. Mol. Cat. B: Enzym*, 66, 177-181, ISSN: 0039-7911.

Jauregui, J., Valderrama, B., Albores, A. & Vazquez-Duhalt, R. (2003). Microsomal transformation of organophosphorus pesticides by white rot fungi. *Biodegradation*, 14, 397–406, ISSN: 0923-9820.

Kahru, A., Pollumaa, L., Reiman, R., Ratsep, A., Liiders, M. & Maloveryan, A. (2000). The toxicity and biodegradability of eight main phenolic compounds characteristic to the oil-shale industry wastewater: a test battery approach. *Environ. Toxicol*, 15, 431–442, ISSN: 1520-4081.

Karim, Z. & Husain, Q. (2009). Guaiacol-mediated oxidative degradation and polymerization of bisphenol A catalyzed by bitter gourd (*Momordica charantia*) peroxidase. *J. Mol. Cat. B: Enzym*, 59, 185–189, ISSN: 0039-7911.

Keharia, H. & Madamvar, D. (2003). Bioremediation concept for treatment of dye containing wastewater: a review. *I. J. Exp. Biol*, 41, 1068–1075, ISSN: 0019-5189.

Kurniawati, S. & Nicell, J. A. (2007). Efficacy of mediators for enhancing the laccase-catalyzed oxidation of aqueous phenol. *Enzym. Microbial. Technol*, 41, 353–361, ISSN: 0141-0229.

Livingston, A. G. (2004). A novel membrane bioreactor for detoxifying industrial wastewater: II. Biodegradation of 3-chloronitrobenzene in an industrially produced wastewater. *Biotechnol. Bioeng*, 41, 427–936, ISSN: 0006-3592.

Lowry, O.H., Rosebrough, N.J., Farr, A.L. & Randall, R.J. (1951). Protein measurement with Folin–phenol reagent. *J. Biol. Chem.* 193, 265–275, ISSN: 0021-9258.

Marco, S.L. & José, A.P. (2007a). Degradation of Reactive Black 5 by Fenton/UV-C and ferrioxalate/H_2O_2/solar light processes. *Dyes and Pigments*, 74, 622-629.

Marco, S.L., Albino, A., Dias Ana Sampaio, Carla, A. & José, A.P. (2007b). Degradation of a textile reactive Azo dye by a combined chemical–biological process: Fenton's reagent-yeast. *Water Res*, 41, 1103-1109, ISSN: 0143-7208.

Matto, M. & Husain, Q. (2009). Decolorization of direct dyes by immobilized turnip peroxidase in batch and continuous processes. *Ecotoxicol. Environ. Saf*, 72, 965-971, ISSN: 0147-6513.

Matto, M. & Husain, Q. (2008). Decolorization of textile effluent by bitter gourd peroxidase immobilized on concanavalin A layered calcium alginate-starch beads. *Journal of Hazardous Materials*, 164, 1540–1546, ISSN: 0304-3894.

Matto, M., & Hussain, Q. (2007). Decolorization of direct dyes by salt fractionated turnip proteins enhanced in the presence of hydrogen peroxide and Redox mediators. *Chemosphere*, 69, 388–345, ISSN: 0045-6535.

Robinson, T., McMullan, G., Marchant, R. & Nigam, P. (2001). Remediation of dyes in textile effluents: a critical review on current treatment technologies with a proposed alternative. *Bioresour. Technol*, 77, 247–255, ISSN: 0960-8524.

Rodriguez-Couto, S., Sanroman, M. & Guebitz, G.M. (2005). Influence of Redox mediators and metal ions on synthetic acid dye decolorization by crude laccase from *T. hirsuta*. *Chemosphere*, 58, 417–422, ISSN: 0045-6535.

Satar, R. & Husain, Q.(2009). Use of Bitter Gourd (*Momordica charantia*) peroxidase together with Redox mediators to decolorize disperse dyes. *Biotechnol. Bioprocess Eng*, 14, 213–219, ISSN: 1226-8372.

Shaffiqu, T.S., Roy, J.J., Nair, R.A. & Abraham, T.E. (2002). Degradation of textile dyes mediated by plant peroxidases. *Appl. Biochem. Biotechnol*, 102/103, 315–326, ISSN: 0273-2289.

Soares, G.M., Costa-Ferreira, M. & Pessoa de Amorim, M.T. (2001a). Decolorization of an anthraquinone-type dye using a laccase formulation. *Bioresour. Technol*, 79, 171–177, ISSN: 0960-8524.

Soares, G.M., Pessoa de Amorim, M.T. & Costa-Ferreira, M. (2001b). Use of laccase together with Redox mediators to decolorize Remazol Brilliant Blue R. *J. Biotechnol*, 89, 123–129, ISSN: 0168-1656.

Sumathi, S. & Manju, B.S. (2000). Uptake of reactive textile dyes by *Aspergillus foetidus*. *Enzym. Microbial. Technol*, 27, 347–355, ISSN: 0141-0229.

Tatsumi, K., Wada, S. & Ichikawa, H. (1996). Removal of chlorophenols from wastewater by immobilized horseradish peroxidase. *Biotechnol. Bioeng*, 51, 126–130, ISSN: 0006-3592.

Torres, E., Bustos-Jaimes, I. & Le Bogne, S. (2003). Potential use of oxidative enzymes for the detoxification of organic pollutants. *Appl. Cat. B: Environ*, 46, 1–15, ISSN: 0926-3373.

Vasantha, L. M. Vuringi, H.B. & Yerramilli, A. (2006). Degradation of azo dye with horse radish peroxidase (HRP). *J. Indian Inst. Sci*, 86, 507–514, ISSN: 0970-4140.

Vazquez-Duarte, H., Garcia-Almendarez, B., Regalado, C. & Whitaker, J. R. (2001). Purification and properties of a neutral peroxidase from turnip (*Brassica napus L var* purple top white globe) roots. *Journal of Agricultural and Food Chemistry*, 49, 4450–4456, ISSN: 0021-8561.

Verma, P. & Madamwar, D. (2002). Decolorization of synthetic textile dyes by lignin peroxidase of Phanerochaete chrysosporium. *Folia Microbiol*, 47, 283–286, ISSN: 0015-5632.

Wada, S., Ichikawa, H. & Tatsumi, K. (1995). Removal of phenols and aromatic amines from wastewater by a combination treatment with tyrosinase and a coagulant. *Biotechnol. Bioeng*, 45, 304–309, ISSN: 0006-3592.

Xu, F., Deussen, H.J.W., Lopez, B., Lam, L. & Li, K. (2001). Enzymatic and electrochemical oxidation of N-hydroxy compounds: Redox potential, electrontransfer kinetics, and radical stability. *Eur. J. Biochem*, 268, 4169–4176, ISSN: 0014-2956.

Zille, A., Tzanov, T., Gubitz, G. M. & Cavaco-Paulo, A. (2003). Immobilized laccase for decolorization of reactive black 5 dyeing effluent. *Biotechnol. Lett*, 25, 1473–1477, ISSN: 0141-5492.

7

Pilot Plant Experiences Using Activated Sludge Treatment Steps for the Biodegradation of Textile Wastewater

Lamia Ayed and Amina Bakhrouf

Laboratoire d'Analyse, Traitement et Valorisation des Polluants de l'Environnement et des Produits, Faculté de Pharmacie, Monastir Tunisie

1. Introduction

Considering both the volume and the effluent composition, the textile industry wastewater is rated as the most polluting among all industrial sectors. Important pollutants are present in textile effluents; they are mainly recalcitrant organics, colour, toxicants and inhibitory compounds (Khelifi et al., 2008).

Textile industries however, have caused serious environmental problems because of the wastewater produced. Most textile industries produce wastewater with relatively high BOD, COD, suspended solids and color. The wastewater may also contain heavy metals depending on the type of coloring substances used. In general, the objective of textile industry wastewater treatment to reduce the level of organic pollutants, heavy metal, suspended solids and color before discharge into the river. Coloring substances are used for dyeing and printing processes. The wastewater from these two processes is the most polluted liquid waste in a textile industry. Biological, chemical, physical or the combination of the three treatment technologies can be used to treat textile industry liquid waste (Suwardiyno and Wenten, 2005). It has been proven that some of these dyes and/or products are carcinogens and mutagens (Manu and Chaudhari 2003). A part from the aesthetic deterioration of the natural water bodies, dyes also cause harm to the flora and fauna in the natural environment (Kornaros and Lyberatos 2006). So, textile wastewater containing dyes must be treated before their discharge into the environment (Forgas et al., 2004). Numerous processes have been proposed for the treatment of coloured waste water e.g., precipitation, flocculation, coagulation, adsorption and wet oxidation (Hongman et al., 2004; Thomas et al., 2006). All these methods have different colour removal capabilities, capital costs and operating speed. Among these methods coagulation and adsorption are the commonly used; however, they create huge amounts of sludge which become a pollutant on its own creating disposal problems (Nyanhongo et al., 2002). Among low cost, viable alternatives, available for effluent treatment and decolourization, the biological systems are recognised, by their capacity to reduce biochemical oxygen demand (BOD) and chemical oxygen demand (COD) by conventional aerobic biodegradation (Forgas et al., 2004; Kornaros and Lyberatos 2006; Balan and

Monteiro, 2001). Work on the use of combined bacterial process to treat textile wastewater has been carried out over the years by many research groups. Recent study has used the combination of anaerobic and aerobic steps in an attempt to achieve not only decolourization but also mineralization of dyes (Forgas et al., 2004; Ong et al., 2005). Aerobic processes have been recently used for the treatment of textile wastewater as standalone processes (Khelifi et al., 2008) and it is confirmed that they are efficient, cost-effective for smaller molecules and that the aerobic reactor is an effective technique to treat industrial wastewater (Coughlin et al., 2002; Coughlin et al., 2003; Buitron et al., 2004; Ge et al., 2004; Sandhaya et al., 2005; Steffan et al., 2005; Sudarjanto et al., 2006). The aerobic reactor has the advantage of being a closed and comparatively homogeneous and stable ecosystem. Since little is known about this ecosystem, a molecular inventory is the first step to describe this dynamic bacterial community without cultivation (Godon et al., 1997). In order to better understand the functions of the bacterial community, a full description of the bacterial ecosystem is required (Bouallagui et al., 2004). Acquisition of DNA sequences is now a fundamental component of most phylogenetic, phylogeographic and molecular ecological studies. Single-strand conformation polymorphism (SSCP) offer a simple, inexpensive and sensitive method for detecting whether or not DNA fragments are identical in sequence, and so can greatly reduce the amount of sequencing necessary (Sunnucks et al., 2000). SSCP can be applied without any a priori information on the species and then can give a more objective view of the bacterial community. SSCP has been applied to study microbial communities from several habits including water, compost and anaerobic digesters (Duthoit et al., 2003; Bouallagui et al., 2004).

In this research, we used the mixture design in the experimental design (Minitab 14.0) to optimize the formulation of the predominant strains isolated from textile waste water plant. After biodegradation, the Chemical Oxygen Demand (COD) and percentage of decolorization were measured. The relationships between the different combinations and products were analyzed by Minitab to select the optimal bacterial combination and to investigate the aerobic degradability of a textile industry wastewater in Tunisia by an aerobic Stirred Bed Reactor (SBR).

2. Materials and methods

2.1 Materials

The microbial strains were microcapsules of *Sphingomonas paucimobilis* ($14×10^7$ cfu), *Bacillus sp.* ($4.2×10^8$ cfu) and *Staphylococcus epidermidis* ($7×10^9$ cfu), which were isolated from textile Waste Water plant in KsarHellal, Tunisia. *Sphingomonas paucimobilis*, *Staphylococcus epidermidis* and *Bacillus sp.* were isolated in previous works of Ayed et al.2009a,b and Ayed et al 2010a,b,c with the ability of degrading azo and triphenylmethane dyes (Congo Red, Methyl Red, Methyl Orange, Malachite Green, Phenol Red, Fushin, Methyl Green and Crystal Violet). All chemicals used were of the highest purity available and of analytical grade.

2.2 Nutrient agar preparation

Nutrient agar was used as the growth medium for microbial isolation. For this purpose, 28 g of nutrient agar was dissolved in 1 l of distilled water, and was then autoclaved at 121 °C for 20 min. After autoclaving, the agar was left to cool at room temperature for 15 min, and it was then poured out into Sterilin © disposable Petri dishes.

2.3 Microbial strain
The culture was cultivated and maintained by weekly transfers on to nutrient agar slants. For production experiments, the culture was revived in nutrient broth (pH 7.0) and freshly prepared 3 h old culture (λ_{600} nm= 1) prepared in Mineral Salt Medium (MSM) at 37 °C, 150 rpm (New Brunswick Scientific Shaker, Edison, NJ) was used as the inoculum. The used medium was composed in 1000 ml of distilled water: glucose (1250 mg/l), yeast extract (3000 mg/l), $MgSO_4$ (100 mg/l); $(NH_4)_2SO_4$ (600 mg/l); NaCl (500 mg/l); K_2HPO_4 (1360 mg/l); $CaCl_2$ (20 mg/l); $MnSO_4$ (1.1 mg/l); $ZnSO_4$ (0.2 mg/l); $CuSO_4$ (0.2 mg/l); $FeSO_4$ (0.14 mg/l) and it was maintained at a constant pH of 7 by the addition of phosphate buffer (Ayed et al., 2010a,b,c).

2.4 Acclimatization
The acclimatization was performed by gradually exposing *Sphingomonas paucimobilis*, *Bacillus sp.* and *Staphylococcus epidermidis* to the higher concentrations of effluent (Kalme et al., 2006). This bacteria were grown for 24 h at 30 °C in 250 ml Erlenmeyer flasks containing in g/l yeast extract (3.0) and glucose (1.25) (pH 7.0). During the investigation, nutrient broth concentration was decreased from 90% (w/v) to 0% (w/v) and finally the organism was provided with effluent as sole source of nutrient. Acclimatization experiments were carried out at optimum temperature (Kalme et al., 2006).

2.5 Operational conditions of laboratory bioreactors
A laboratory scale aerobic bioprocess was used in this study. The aerobic system used was SBR bioreactor. The system was operated continuously at a constant temperature of 30 °C using an external water bath. A continuous stirred tank reactor with a 500 ml working volume was used. Mixing was assured by the continuous rotation of the magnetic stirrer. The system was first inoculated with a microbial consortia (*Sphingomonas paucimobilis*, *Bacillus sp.* and *Staphylococcus epidermidis*) obtained from a textile wastewater treatment plant. These inocula were selected because of the large variety of microorganisms that could be found in the biomass degrading dyes in textile wastewater, and because mixed cultures offer considerable advantages over the use of pure culture. In fact, individual strains may attack the dye molecules at different position or may use decomposition products produced by another strains for further decomposition. In fact, it is mentioned that adaptation is important for successful decolorization, and as acclimation occurred, the decolorization time becomes constant (Buitron and Quezada Moreno, 2004). The system was fed by a peristaltic pump with the textile effluent obtained from textile wastewater plant in Ksar Hellal (Tunisia), and its pH was maintained at approximately 7. Air was provided from the bottom of the aeration of the combined bacterial process using diffusers and an air pump. Bioreactors operating conditions were (COD: 1700 (mg O_2/l); BOD_5: 400 (mg O_2/l); Color : 3600 (U.C); pH: 7; MES: 810 (mg/l)).

2.6 Analytical methods
The effluent from each bioreactor was collected daily, centrifuged at 6000 rpm for 10 min and analysed for color, COD, pH, volatile suspended solids (VSS) and colonies forming units (cfu). COD and color measurements were carried out on the clear supernatant. Color was measured by an UV–vis spectrophotometer (Spectro UV-Vis Double Beam PC Scnning spectrophotomètre UVD-2960) at a wavelength of 275 nm in which maximum absorbance spectra was obtained. The decolorization and COD removal were calculated according to the following formulation (Eq 1and Eq 2) (Ayed et al., 2009a,b).

In this study, *Sphingomonas paucimobilis*, *Bacillus sp.* and Filamentous bacteria were used as mixture starters, with different proportions ranging from 0 to 100%, as shown in Table 1. Decolorization experiments were taken according to the ratio given by the experimental design, and 10% of mixed culture were inoculated into the effluent (3.0 g/l yeast extract and 1.25 g/l glucose) at 37°C for 10 h in shaking conditions (150 rpm) (Ayed et al., 2010a,b,c).

$$\% \text{ Decolorization } = \frac{(I - F)}{I} \times 100 \tag{1}$$

Where I was the initial absorbance and F the absorbance at incubation time t

$$\text{COD removal } (\%) = \frac{\text{initial COD}(0 \text{ h}) - \text{observed COD}(t)}{\text{initial COD}(0 \text{ h})} \times 100 \tag{2}$$

The pH was measured using a digital calibrated pH-meter (Inolab, D-82362 Weilheim Germany). All assays were carried in triplicate.

Assay	Sphingomonas paucimobilis	Bacillus sp.	Staphylococcus epidermidis	Total	COD Removal (%)	Decolorization (%)
1	0.66667	0.16667	0 .16667	1.00000	60	70
2	0.50000	0.50000	0.00000	1.00000	76	77
3	0.50000	0.00000	0.50000	1.00000	77	88
4	0.33333	0.33333	0.33333	1.00000	75	80
5	0.00000	1.00000	0.00000	1.00000	70	77
6	1.00000	0.00000	0.00000	1.00000	81	89
7	0.16667	0.16667	0.66667	1.00000	53	63
8	0.00000	0.00000	1.00000	1.00000	49	55
9	0.00000	0.50000	0.50000	1.00000	45	44
10	0.16667	0.66667	0.16667	1.00000	60	64

Table 1. Mixture design matrix with the experimental analysis

2.7 Pilot plant design
As described earlier, the pilot plant comprised several treatment steps.

3. Results and discussion

3.1 Model establishment
Through linear regression fitting, the regression models of tow responses (COD % and decolorization %) were established. The regression model equations are as follows:

$$Y_{\text{decolorization}\%} = 85.34 \, S1 + 67.70 \, S2 + 55.43 \, S3 + (-21.77) \, S1*S2 + (67.69) \, S1*S3 + (-73.59) \, S2*S3$$

$$R^2 = 84.82\%; \, P=0.09$$

$$Y_{\text{COD}\%} = 77.11 \, S1 + 70.11 \, S2 + 49.02 \, S3 + (-1.55) \, (S1*S2) + (44.27) \, (S1*S3) + (-57.73) \, (S2*S3)$$

$$R^2 = 75.27\%; \, P = 0.2$$

Where S1: Sphingomonas paucimobilis; S2: Bacillus sp. and S3: Staphylococcus epidermidis

3.2 Effect of formulation on the percentage of decolorization and COD removal of effluent

The mixture design is now used widely in the formulation experiment of food, chemicals, fertilizer, pesticides, and other products. It can estimate the relationship between formulation and performance through regression analysis in fewer experiment times (Zhang et al., 2006).

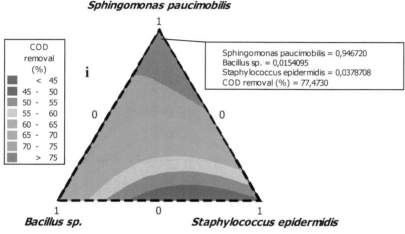

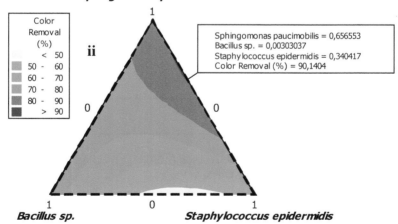

Fig. 1. Mixture contour plots between the variables (*Sphingomonas paucimobilis, Bacillus sp* and *Staphylococcus epidermidis.*) for i COD removal (%), ii Color removal (%).

Zhang et al. (2006) studied the formulation of plant protein beverage using the mixture design, obtaining the optimized combination of walnut milk, peanut milk, and soy milk. In the mixture design, the effect of the change of variables on the responses can be observed on the ternary contour map. Figure 1 shows the effect of the interaction of *Sphingomonas paucimobilis*, *Bacillus sp.* and *Staphylococcus epidermidis* on the decolorization of effluent; Figure 1 shows the effect of the interaction of *Sphingomonas paucimobilis, Bacillus sp.* and *Staphylococcus epidermidis* on the variation of COD. The statistical significance of the ratio of mean square variation due to regression and mean square residual error was tested using analysis of variance. ANOVA is a statistical technique which subdivides the total variation in a set of data into component parts associated with specific sources of variation for the purpose of testing hypotheses on the parameters of the model.

Only results obtained for decolorization and COD removal were presented herein for clarity of purpose. According to the ANOVA (Table 2 and 3), the regression adjusted average squares were (305.8) and (231), the linear regression adjusted average squares were (1529.3) and (1115.02) allowed the calculation of the Fisher ratios (F-value) for assessing the statistical significance. The model F-value (4.33) and (2.43) implies that most of the variation in the response can be explained by the regression equation.

Source	Degrees of freedom	Sum of square	Sum of adjusted squares	adjusted average squares	F-ratio	P-value (significance)
Regression	5	1529,3	1529,303	305,861	4,33	0,090
Linear regression	2	996,33	521,279	260,639	3,69	0,123
Quadratic regression	3	532,97	532,970	177,657	2,52	0,197
Residual error	4	282,30	282,297	70,574		
Total	9	1811,60				

Table 2. Analysis of variance of % decolorization (ANOVA) for the selected linear and interactions model for effluent textile wastewater

The P-value for the regression obtained R^2= 84.82%; P=0.09 for decolorization was less than 0.1 and means consequently that at least one of the term in the regression equation has significant correlation with the response variable.

The associated P-value is used to judge whether F-ratio is large enough to indicate statistical significance. A P-value is more than 0.1 (i.e. α =0.05 or 95% confidence) indicates that the model is not to be considered statistically significant. The non-significant value of lack of fit (>0.05) revealed that the quadratic model is statistically significant for the response and therefore it can be used for further analysis (Zhou et al., 2007). The ANOVA test also shows a term for residual error, which measures the amount of variation in the response data left unexplained by the model (Xudong and Rong, 2008).The collected data were analyzed by using Minitab® 14 Statistical Software for the evaluation of the effect of each parameter on the optimization criteria. In order to determine the effective parameters and their confidence levels on the color removal process, an analysis of variance was performed. A statistical analysis of variance (ANOVA) was performed to see which process parameters were statistically significant. F-test is a tool to see which process parameters have a significant effect on the dye removal value. The F-value for each process parameter is simply a ratio of the mean of the squared deviations to the mean of the squared error. The color removal from the real textile wastewater was investigated in different experimental conditions.

Source	Degrees of freedom	Sum of square	Sum of adjusted squares	adjusted average squares	F-ratio	P-value (significance)
Regression	5	1155,02	1155,023	231,005	2,43	0,205
Linear regression	2	885,44	470,406	235,203	2,48	0,199
Quadratic regression	3	269,58	269,578	89,859	0,95	0,498
Residual error	4	379,48	379,477	94,869		
Total	9	1534,50				

Table 3. Analysis of variance of COD% (ANOVA) for the selected linear and interactions model for effluent textile wastewater

The mixture surface plots (Figure 2), which are a three-dimensional graph, was represented using COD and color removal were represented based on the simultaneous variation of *Sphingomonas paucimobilis*, *Bacillus sp.* and *Staphylococcus epidermidis* in the consortium composition ranging from 0 to 100 % for each strain. The mixture surface plot also describing individual and cumulative effect of these three variables and their subsequent effect on the response (Liu et al., 2009; Ayed et al., 2010b,c).

The mixture contour plots between the variables such as *Sphingomonas paucimobilis*, *Bacillus sp.* and *Staphylococcus epidermidis* are given in Figure 2. The lines of contour plots predict the values of each response at different proportion of *Sphingomonas paucimobilis*, *Bacillus sp.* and *Staphylococcus epidermidis*. These values are more or less same to the experimental values.

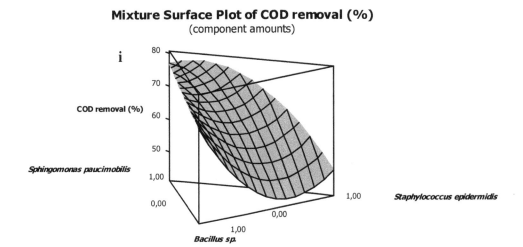

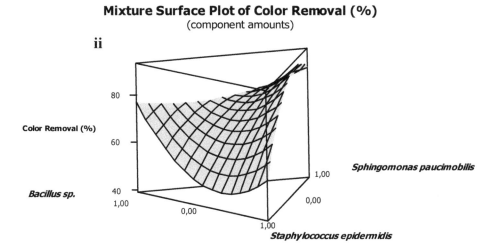

Fig. 2. Mixture surface plots between the variables (*Sphingomonas paucimobilis, Bacillus sp* and *Staphylococcus epidermidis.*) for i COD removal (%), ii Color removal (%).

4. Conclusions

The developed consortium showed a better decolorization yields as compared to pure cultures, which proved a complementary interaction among various isolated bacteria. The consortium achieved significantly a higher reduction in color (90.14%) and COD removal (77.47%) in less time (96h). The biodegradation of the effluent textile wastewater was achieved by the developed consortium using *Sphingomonas paucimobilis, Bacillus sp.* and *Staphylococcus epidermidis.*

5. References

Bouallagui, H., Torrijos, M., Godon, J.-J., Moletta, R., Ben-Cheik, R., Touhami, Y., Delegenes, J.-P., Hamdi, M., 2004. Microbial monitoring by molecular tools of a two-phase anaerobic bioreactor treating fruit and vegetable wastes. Biotechnol. Lett. 26, 857–862.

Buitron, G., Quezada, M., Moreno, G., 2004. Aerobic degradation of the azo dye acid red 151 in a sequencing batch biofilter. Bioresour. Technol. 92, 143–149.

Duthoit, F., Godon, J.-J., Montel, M.-C., 2003. Bacterial community dynamics during production of registered designation of origin salers cheese as evaluated by 16S rRNA gene single-strand conformation polymorphism analysis. Appl. Environ. Microbiol. 69, 3840–3848.

Khelifi, E., Gannoun, H., Touhami, Y., Bouallagui, H., Hamdi, M., 2008. Aerobic decolourization of the indigo dye-containing textile wastewater using continuous combined bioreactors. J. Hazard. Mater. 152, 683–689.

Kornaros, M., Lyberatos, G., 2006. Biological treatment of wastewaters from a dye manufacturing company using a trickling filter. J. Hazard. Mater. 136, 95–102.

Manu, B., Chaudhari, S., 2003. Decolourization of indigo and azo dyes in semi continuous reactors with long hydraulic retention time. Process Biochem. 38, 1213–1221.

Forgas, E., Cserhati, T., Oros, G., 2004. Removal of synthetic dyes wastewaters: a review. Environ. Int. 30, 953–971.

Godon, J.-J., Zumstein, E., Darbert, P., Habouzit, F., Moletta, R., 1997. Molecular bacterial diversity of an anaerobic digester as determined by small-subunit rDNA sequence analysis. Appl. Environ. Microbiol. 63, 2802–2813.

Hongman, H., Jiti, Z., Jing, W., Cuihong, D., Bin, Y., 2004. Enhancement of laccase production by P. ostreatus and its use for the decolourization of anthraquinone dye. Process Biochem. 39, 1415–1419.

Nyanhongo, G.S., Gomes, J., Gubitz, G.M., Zvauya, R., Read, J., Steiner, W., 2002. Decolorization of textile dyes by laccases froma newly isolated strain of Trametes modesta. Water Res. 36, 1449–1456.

Khelifi, E., Gannoun, H., Touhami, Y., Bouallagui, H., Hamdi, M., 2008. Aerobic decolourization of the indigo dye-containing textile wastewater using continuous combined bioreactors. J. Hazard. Mater. 152, 683–689.

Ge, Y., Yan, L., Qinge, K., 2004. Effect of environment factors on dye decolourization by P. sordida ATCC90872 in an aerated reactor. Process Biochem. 39, 1401–1405.

Coughlin, M.F., Kinkle, B.K., Bishop, P.L., 2002. Degradation of acid orange 7 in an aerobic biofilm. Chemosphere 46, 11–19.

Coughlin, M.F., Kinkle, B.K., Bishop, P.L., 2003. High performance degradation of azo dye acid orange 7 and sulfanilic acid in a laboratory scale reactor after seeding with cultured bacterial strains. Water Res. 37, 2757–2763.

Sandhaya, S., Padmavathy, S., Swaminathan, K., Subrahmanyam, Y.V., Kaul, S.N., 2005. Microaerophilic-aerobic sequential batch reactor for treatment of azo dyes containing simulated wastewater. Process Biochem. 40, 885–890.

Steffan, S., Badri, L., Marzona, M., 2005. Azo dye biodegradation by bacterial cultures immobilized in alginate beads. Environ. Int. 31, 201–205.

Sudarjanto, G., Lehmann, B.K., Keller, J., 2006. Optimization of integrated chemical-biological degradation of a reactive azo dye using response surface methodology. J. Hazard. Mater. B138, 160–168.

Sunnucks, P., Wilson, A.C.C., Beheregaray, L.B., Zenger, K., French, J., Taylor, A.C., 2000. SSCP is not so difficult: the application and utility of single stranded conformation polymorphism in evolutionary biology and molecular ecology. Mol. Ecol. 9, 1699–1710.

Suwardiyono,T.S , Wenten,I.G.. (2005) Treatment of Textile Wastewater by a Coupling of Activated Sludge Process with Membrane Separation Journal of Water and Environment Technology, 3,125-136

Thomas, B., Aurora, T., Wolfgang, S., 2006. Electrochemical decolourization of dispersed indigo on boron-doped diamond anodes. Diamond Relat. Mater. 15, 1513–1519.

Permissions

The contributors of this book come from diverse backgrounds, making this book a truly international effort. This book will bring forth new frontiers with its revolutionizing research information and detailed analysis of the nascent developments around the world.

We would like to thank Prof. Peter J. Hauser, for lending his expertise to make the book truly unique. He has played a crucial role in the development of this book. Without his invaluable contribution this book wouldn't have been possible. He has made vital efforts to compile up to date information on the varied aspects of this subject to make this book a valuable addition to the collection of many professionals and students.

This book was conceptualized with the vision of imparting up-to-date information and advanced data in this field. To ensure the same, a matchless editorial board was set up. Every individual on the board went through rigorous rounds of assessment to prove their worth. After which they invested a large part of their time researching and compiling the most relevant data for our readers. Conferences and sessions were held from time to time between the editorial board and the contributing authors to present the data in the most comprehensible form. The editorial team has worked tirelessly to provide valuable and valid information to help people across the globe.

Every chapter published in this book has been scrutinized by our experts. Their significance has been extensively debated. The topics covered herein carry significant findings which will fuel the growth of the discipline. They may even be implemented as practical applications or may be referred to as a beginning point for another development. Chapters in this book were first published by InTech; hereby published with permission under the Creative Commons Attribution License or equivalent.

The editorial board has been involved in producing this book since its inception. They have spent rigorous hours researching and exploring the diverse topics which have resulted in the successful publishing of this book. They have passed on their knowledge of decades through this book. To expedite this challenging task, the publisher supported the team at every step. A small team of assistant editors was also appointed to further simplify the editing procedure and attain best results for the readers.

Our editorial team has been hand-picked from every corner of the world. Their multi-ethnicity adds dynamic inputs to the discussions which result in innovative outcomes. These outcomes are then further discussed with the researchers and contributors who give their valuable feedback and opinion regarding the same. The feedback is then collaborated with the researches and they are edited in a comprehensive manner to aid the understanding of the subject.

Apart from the editorial board, the designing team has also invested a significant amount of their time in understanding the subject and creating the most relevant covers. They scrutinized every image to scout for the most suitable representation of the subject and create an appropriate cover for the book.

The publishing team has been involved in this book since its early stages. They were actively engaged in every process, be it collecting the data, connecting with the contributors or procuring relevant information. The team has been an ardent support to the editorial, designing and production team. Their endless efforts to recruit the best for this project, has resulted in the accomplishment of this book. They are a veteran in the field of academics and their pool of knowledge is as vast as their experience in printing. Their expertise and guidance has proved useful at every step. Their uncompromising quality standards have made this book an exceptional effort. Their encouragement from time to time has been an inspiration for everyone.

The publisher and the editorial board hope that this book will prove to be a valuable piece of knowledge for researchers, students, practitioners and scholars across the globe.

List of Contributors

Taner Yonar
Uludag University, Environmental Engineering Department, Gorukle, Bursa, Turkey

Zongping Wang, Miaomiao Xue, Kai Huang and Zizheng Liu
Huazhong University of Science and Technology, China

Falah Hassan Hussein
Chemistry Department, College of Science, Babylon University, Iraq

Idil Arslan-Alaton and Tugba Olmez-Hanci
Istanbul Technical University, Turkey

Farah Maria Drumond Chequer and Danielle Palma de Oliveira
USP, Departamento de Análises Clínicas, Toxicológicas e Bromatológicas, Faculdade de Ciências Farmacêuticas de Ribeirão Preto, Universidade de São Paulo, Ribeirão Preto – SP, Brazil

Daniel Junqueira Dorta
USP, Departamento de Química, Faculdade de Filosofia, Ciências e Letras de Ribeirão Preto, Universidade de São Paulo, Ribeirão Preto – SP, Brazil

Farrukh Jamal
Department of Biochemistry, Ram Manohar Lohia Avadh University, Faizabad, U.P., India

Lamia Ayed and Amina Bakhrouf
Laboratoire d'Analyse, Traitement et Valorisation des Polluants de l'Environnement et des Produits, Faculté de Pharmacie, Monastir Tunisie